Quantum Adaptivity in Biology: From Genetics to Cognition

Masanari Asano · Andrei Khrennikov
Masanori Ohya · Yoshiharu Tanaka
Ichiro Yamato

Quantum Adaptivity in Biology: From Genetics to Cognition

Springer

Masanari Asano
Liberal Arts Division
Tokuyama College of Technology
Tokuyama
Japan

Andrei Khrennikov
International Center for Mathematical Modeling in Physics and Cognitive Sciences
Linnaeus University
Växjö
Sweden

Masanori Ohya
Information Science
Tokyo University of Science
Tokyo
Japan

Yoshiharu Tanaka
Information Science
Tokyo University of Science
Tokyo
Japan

Ichiro Yamato
Department of Biological Science and Technology
Tokyo University of Science
Tokyo
Japan

ISBN 978-94-024-0386-2 ISBN 978-94-017-9819-8 (eBook)
DOI 10.1007/978-94-017-9819-8

Springer Dordrecht Heidelberg New York London

Softcover reprint of the hardcover 1st edition 2015

Printed on acid-free paper

Springer Science+Business Media B.V. Dordrecht is part of Springer Science+Business Media (www.springer.com)

Science is beautiful when it makes simple explanations of phenomena or connections between different observations. Examples include the double helix in biology and the fundamental equations of physics.

Stephen Hawking

I'm fascinated by the idea that genetics is digital. A gene is a long sequence of coded letters, like computer information. Modern biology is becoming very much a branch of information technology.

Richard Dawkins

Foreword

Unlike any other discipline in the natural sciences, Biology has benefitted tremendously from intriguing ideas and novel concepts from outside the subject's area throughout the last century. The apparently distinct topics of chemistry, physics, mathematics and informatics became integral and indispensable matters of biological research that blended surprisingly well with organismal studies of the last centuries' Biology. In this very aspect, Biology is reminiscent of the principal ideas of ancient Philosophy, as both fields specialize in their quest for understanding the essence of 'life', 'meaning' and 'truth'.

As an inherent consequence of organismal plasticity and diversity, 'truth' in biological findings is given support by studying the probability of a discrete or gradual trait in a population, which uses stochastic expressions of classical mathematics. Here, again, striking similarities between biology and philosophy exist: While modern deductive biology uses mathematics to describe processes as precisely as possible such as the dynamics of chemical reactions inside a cell or even growth and development of organs, pure mathematical formalism dominated philosophy and ruled deductive reasoning for almost two millennia.

Established by Aristotle's logic, especially in his theory of syllogism about 300 BC, the meaning of words and thoughts can be expressed in common mathematical formalism to deliver significance. This pure logic is bijective and unbiased from feelings or subjective observations. Besides its influence on natural sciences, Aristotle's formal logic and the resulting Aristotelian philosophy had a great impact on theology, especially Islamic and Christian religion. In addition, he introduced deductive reasoning also into his own biological studies and, hence, Aristotle can be considered as the founder of modern deductive biology. He was even ambitious enough to adopt his theory of pure logic to the operational processes in the brain.

Ironically, the period of Aristotle's logic formalism faced an end during the epoch of Enlightenment that climaxed in Charles Darwin's and Alfred R. Wallace's biological ideas of speciation and evolution!

During a period of almost 200 years, theories of modern mathematical logic and Aristotelian logic were seemingly incompatible. In recent years, however, it is stated that modern mathematical formalism and Aristotle's theories of logic disclose striking similarities. Moreover, biological conceptions merged with these ideas and formed the current basis of *modern biology that superficially splits into overlapping, but yet distinct disciplines such as bioethics, biochemistry, biophysics, biostatistics or bioinformatics*.

Modern biologists still aim at finding scientific ‘truth’, at identifying how ‘life’ functions or how nerve cells compute ‘meaning’. Intriguingly, with high-throughput methods at hand, biologists accumulate an enormous amount of data within a short period of time that gives a renaissance to descriptive biology of pre-Aristotelian reasoning or of the epoch of Enlightenment. Methods of classical probability are imposed to mine all kinds of observables for statistical significance, but a large amount of information remains cryptic.

Especially, next generation sequencing technologies provide strings of genomic DNA-sequences that encode for all the genetic make-up that determines an individual or a species at extraordinary speed. As a consequence, an overwhelming volume of genomic sequences is generated that await detailed analysis by bioinformatics. This large amount of information produces huge problems in data storage and evaluation that might benefit from novel ground-breaking ideas or progress in technology.

The present monograph by Masanari Asano, Andrei Khrennikov, Masanori Ohya, Yoshiharu Tanaka and Ichiro Yamato pursues such novel ground-breaking ideas in approaching phenomena of modern biology by quantum probabilistic formalism. Unlike the description of quantum physical processes at molecular scale, e.g. the processes of exciton transfer in photosynthesis or fluorescence resonance energy transfer between different molecules, the authors use the *operational formalism* of quantum theory to address biological problems at diverse macroscopic biological scales. They discuss and describe the application of quantum-like probability to various biological examples such as sequence and gene expression analysis, bacterial growth or epigenetics studies as well as cognitive science.

The book highlights the similarities between mathematical formalism of quantum probability in physics and statistical experimental data by quantum bio-informatics. The authors carefully discuss current limitations of the general quantum information theory and propose that an extended formalism might be required to suit special biological problems. More importantly, however, they convincingly explain that some biological observations violate classical mathematic formalism, which might successfully be addressed by quantum-like probability or quantum bioinformatics.

This monograph emphasizes the great potential of quantum-like probability in life science, especially for the many dynamical processes in biological studies. While classical stochastic formalism is rather static, quantum probability is more advanced and allows the representation of dynamical biological processes as different states in a framework of quantum fluctuations.

Finally, the authors provide a vital debate about how evolutionary aspects can possibly be described by quantum-like models. Therefore, it is not surprising that the authors address how quantum-like formalism and probability can support biological research in clarifying what ‘life’, ‘meaning’ and ‘truth’ is. Following

Aristotelian logic, they concluded that quantum-like models hold the potential for *unifying Neo-Darwinism and Neo-Lamarckism theories* in modern deductive biology.

Although this book does not solve the problems of modern biological data analysis, it constitutes an extraordinary attempt to show the applicability of quantum theory and quantum-like formalism to macroscopic observables, which is novel and a very stimulating read.

Dierk Wanke
Saarland University and University of Tübingen
Germany

Preface

The aim of this book is to introduce a theoretical/conceptual principle (based on *quantum information theory* and *non-Kolmogorov probability theory*) to understand information processing phenomena in biology as a whole—the *information biology*—a new research field, which is based on the application of open quantum systems (and, more generally, *adaptive dynamics* [173, 26, 175]) outside of physics as a powerful tool. Thus this book is about information processing performed by bio-systems. Since quantum information theory generalizes classical information theory and presents the most general mathematical formalism for the representation of information flows, we use this formalism. In short, this book is about *quantum bio-information*. However, it is not about quantum physical processes in bio-systems. We apply the mathematical formalism of quantum information as an operational formalism to bio-systems at all scales: from genomes, cells, and proteins to cognitive and even social systems.

F. Crick proposed a central dogma in molecular biology to understand the genetic coding problem in biology in 1970 [62]. So far researchers in biological sciences have elucidated individual molecular mechanisms of any information processing phenomena in biology, such as signal transduction, differentiation, cognition and even evolution. However, we do not have basic/unified principles underlying such information flows in biology from genes to proteins, to cells, to organisms, to ecological systems and even to human social systems. It seems that we are now at the brink of the crisis (catastrophe) of the integrity of our earth system including human societies. We are longing for any possible tools to predict our state in the future.

Nowadays, quantum information theory is widely applied for microscopic physical carriers of information such as photons and ions. This is definitely one of the most rapidly developing domains of physics. Classical information theory is based on a special model of probability theory, namely the model proposed by A.N. Kolmogorov in 1933 [148]. Quantum information theory is based on the quantum probability model—the calculus of complex probability amplitudes. The latter was elaborated by the founders of quantum mechanics, first of all, by M. Born and J. von Neumann. Quantum probabilities exhibit many unusual (even exotic and mystical) features. In particular, they violate the main laws of classical *Kolmogorovian probability*. As was emphasized by R. Feynman (one of the physical genies of the twentieth century), the quantum interference phenomenon demonstrated in the two slit experiment (where a quantum particle interferes with itself) can be probabilistically interpreted as *violation of the formula of total probability*. The latter is one of the most fundamental laws of classical probability theory, the heart of *Bayesian analysis* with corresponding applications to the game theory and decision making. Thus quantum physics has demonstrated that the classical probability model (as any mathematical model) has a restricted domain of application. In particular, it can be used in classical statistical mechanics, but not in quantum mechanics.

The situation in probability theory is similar to the situation in geometry. During two thousand years Euclidean geometry was considered as the only possible mathematical model of physical space. (We recall that E. Kant even claimed that this geometry was one of the basic elements of reality.) However, the discovery of N. Lobachevsky showed that other consistent mathematical geometric models were possible as well. Later, Riemann by inventing geometric models known nowadays as Riemann manifolds opened the doors to a variety of geometric worlds. Finally, the genial mind of A. Einstein coupled these mathematical models of geometry to the physical reality by developing the theory of general relativity. (And Lobachesvky's geometry has applications in special relativity).

Physics was one of the first scientific disciplines that were mathematically formalized. Plenty of mathematical theories, which were born in physical studies, have later found applications in other domains of science. The best example is the differential calculus originally developed by Newton for purely physical applications, but nowadays is applied everywhere, from biology to finances. A natural question arises: *Can quantum and, more generally, non-Kolmogorov probability be applied anywhere besides quantum physics?* In this book we shall demonstrate that the answer is positive and that biology is a novel and extended field for such applications. We noticed that biological phenomena, from molecular biology to cognition, often violate classical total probability conservation law [26]. There is plenty of corresponding experimental statistical data.[1] Therefore, it is natural to apply non-Kolmogorovian probability to biology (in the same way as non-Euclidean geometric models are applied in physics). Since the quantum probabilistic model is the most elaborated among non-Kolmogorovian models, it is natural to start with quantum probabilistic modelling of biological phenomena. However, since biological phenomena have their own distinguishing features, one can expect that the standard quantum formalism need not match completely with biological applications. Novel generalizations of this formalism may be required. And this is really the case, see Chap. 4.

[1]A lot of data has been collected in cognitive science and psychology; unfortunately, in molecular biology we have just a few experimental data collections which can be used for comparison of classical and nonclassical probabilistic models. We hope that the present book will stimulate corresponding experimental research in molecular biology.

In this book we show that quantum theory considered as an *operational formalism* can be applied to any biological scale by representing statistical experimental data by means of quantum probability and information. In the last years a general model representing all basic information flows in biology (from molecular biology to cognitive science and psychology and to evolution) on the basis of quantum information theory has been elaborated. In this book the general scheme of embedding of biological information processing into the quantum information formalism is presented and the foundational issues related to usage of quantum representation for macroscopic bio-systems, such as genome, protein,..., cognitive system,..., bio-population are discussed in detail and clarified.

Since a biological system is *fundamentally open*, i.e. it cannot survive without contact with an environment, it is natural to apply the powerful and well-developed apparatus of *theory of open quantum systems* and more generally *adaptive dynamics* to the description of biological information flows. For those who have some preliminary knowledge of quantum mechanics, we say that it seems that Schrödinger's dynamics (describing the evolution of isolated systems in quantum physics) seems to be not so much useful for biological applications. One has to use quantum master equation and its generalizations (see again Chap. 4).

The "constructive know-how" of our quantum bio-information is the procedure of reconstruction of quantum(-like) operators on the basis of the experimental statistical data. *Such an operational representation can be used to predict probabilistic results of new experiments.* Thus quantum bio-information is not only a powerful descriptive formalism unifying the variety of biological processes, but it also has *predictive power*.

Quantum bio-information has already been applied to such basic phenomena as cell differentiation, diauxie (two phase) growth of *Escherichia coli* in glucose and lactose mixed medium, irrational behaviour in games of the Prisoner's Dilemma type, non-Bayesian probability updating in cognitive psychology and evolution theory (unifying Darwinism and Lamarckism and supporting recent theoretical studies in epigenetic). On this basis, we believe that quantum bio-information is the most predictive tool to know our future state on earth. We expect that this quantum operator formalism is a kind of a brave trial to unify our social and natural sciences. We now ask many researchers to recognize the usefulness of this formalism to understand any information processing in biology from micro- to macro-scale and to make benefit for human beings and society on earth.

This approach raises the deep foundational problem: application of information laws of quantum mechanics to *macro bio-systems*. We hope that our book may generate a foundational debate on the approach of experts from physics and biology (in the latter, in a very general sense: from molecular biology to social science). Our approach to biology also raises deep philosophic questions about the role of information in living and physical systems, see Chap. 1.

This book does not present a complete quantum-information account of biological phenomena. Nevertheless, some important examples of applications are given, see Chaps. 5, 6 and 8.

This book is intended for diverse groups of readers: biologists (molecular biology, especially genetics and epigenetics), experts in cognitive science, decision making, and sociology, psychologists, physicists and mathematicians working on problems of quantum probability and information, experts in quantum foundations (physicists and philosophers).

We start the book with an introduction followed by two chapters devoted to fundamentals: Chapter 2 on classical and quantum probability (it also contains a brief introduction to quantum formalism) and Chap. 3 on information approach to molecular biology, genetics and epigenetics. The latter is basic for proceeding to applications of quantum(-like) theory to molecular biology, see Chap. 5. On the other hand, Chap. 3 might be difficult for experts in physics, mathematics, cognition and psychology. Therefore, those who are not interested in applications of quantum methods to molecular biology can jump directly to Chap. 6. However, a part of the biological fundamentals presented in Chap. 3 is used in Chap. 8, which is about the application of open quantum systems to epigenetic evolution.

Tokyo, 2008–2014
Växjö
Moscow

Masanari Asano
Andrei Khrennikov
Masanori Ohya
Yoshiharu Tanaka
Ichiro Yamato

Acknowledgments

We are grateful to Dr. T. Ando, School of Biology, Georgia Institute of Technology, 250 14th Street NW, Atlanta, Georgia 30318-5304, USA for his critical reading of this chapter and for valuable discussion and comments to improve it.

We are also grateful to Dr. I. Basieva, International Center for Mathematical Modeling in Physics and Cognitive Science, Linnaeus University, Sweden, for her important contributions to our joint works on applications of theory of open quantum systems to cognitive science, psychology, and molecular biology and Chaps. 1, 5, 6, 10.

We also want to thank all who participated in discussions on applications of the formalism of quantum information theory to biology and cognitive science, especially, L. Accardi, H. Atmanspacher and T. Filk, J. Busemeyer, E. Dzhafarov, T. Hida, A. Lambert-Mogiliansky, K.H. Fichtner and L. Fichtner, W. Freudenberg, E. Haven, E. Pothos, N. Watanabe, T. Matsuoka, K. Sato, S. Iriyama, T. Hara. And we thank T. Hashimoto and Y. Yamamori for supporting the experiments on optical illusion and *E. coli*'s glucose preference.

The basic results of the book were obtained on the basis of the projects Quantum Bio-Informatics, Tokyo University of Science, and Modeling of Complex Hierarchic systems, Linnaeus University.

Contents

Chapter 1
Introduction

Abstract We present motivations for applications of the formalism of quantum theory to biology. This formalism is interpreted as an operational formalism for description of data of the biological origin (from genetics and cellular biology to cognition): states of bio-systems are represented by vectors in complex linear space, observables by linear operators, probabilistic interpretation is borrowed from quantum mechanics. Irreducible contextuality of bio-systems is an important source of quantum representation of information about behavior and dynamics of these systems. We compare biological and quantum contextualities. Adaptivity to concrete contexts (in particular, the environment) plays an important role in evolution of bio-systems. In quantum theory adaptivity is mathematically represented with the aid of theory of open quantum systems and its generalizations. We discuss the possibility to apply quantum adaptive dynamics to biology. We show that contextuality and adaptivity of bio-systems leads to violation of laws of classical probability theory (based on the Kolmogorov axiomatics, 1933). Finally, we emphasize the role of quantum information in life science.

Keywords Complexity of biological and quantum systems · Operational description of behavior and dynamics · Contextuality · Adaptive dynamics · Open quantum systems · Quantum information theory · Classical and quantum probability

1.1 Complexity of Information Processing in Biological Systems

Biological systems are incredibly complex processors of information. Of course, the human brain is the champion in the community of biological information processors. However, even a single cell has extremely complicated information life. Its genome is a huge information structure. And nowadays we know that the cell's information capacity is not reduced to its genome. Cell's epigenome plays a crucial role in the

M. Asano et al., *Quantum Adaptivity in Biology: From Genetics to Cognition*,
DOI 10.1007/978-94-017-9819-8_1

determination of levels of genes' expressions. And, from the viewpoint of information theory, epigenome is even more complicated structure than genome itself.[1]

To describe a complex information system is a very difficult task. One has to find and describe properly all *intrinsic variables* involved in the information flows, constraints coupling different sets of these variables, and construct a dynamical system for temporal behavior of these variables. In reality the situation is much more complicated, because biological systems are *probabilistic information processors.* "Decisions" taken by biological systems (from cells to brains) are not deterministic. In the process of evolution they have learnt that it is more profitable to react quickly by using probabilistic models of decision making than to analyze *context* completely and to proceed deterministically. Hence, besides the aforementioned description of variables and deterministic interrelations between them, a proper model of a biological decision maker[2] also has to reflect all possible probabilistic correlations between these variables. Such a model has to be completed by a random dynamical system, which gives a temporal picture of behavior of these variables.

Sometimes (for simple situations) it is possible to realize the presented program and to create a detailed model of information flows. However, even for a single cell, this is very difficult. Therefore scientists create rough descriptions, for instance, by using coarse graining of the intrinsic variables. Finally, one may come to conclusion that the best research strategy is to restrict modeling to variables determining quantities, which can be measured (e.g., levels of genes' expressions, levels of inherited epimutations, frequencies of neurons' firings,...). However, even this minimalist program is not trivial at all. Measurable quantities can depend on the intrinsic variables in a very complex way and it would be practically impossible to establish functional relations of the form

$$A = A(x_1, ..., x_N), \tag{1.1}$$

where A is a measurable quantity and $x_1, ..., x_N$ are intrinsic variables. In such a situation one would give up attempts to connect the intrinsic variables with measurable quantities and to realize this is the first step towards the usage of *quantum formalism.* One of the new approaches to the above problems is based on the adaptive dynamics, which will be discussed in Chap. 4.

---

[1] Sometimes genome is compared with a computer's program and epigenome with a (huge) set of instructions on the usage of this software, see Chap. 8 for details.

[2] We treat the process of decision making in a very general setup. For example, in our terminology the *E-coli* bacteria "decide" whether to consume lactose or not in the presence or absence of glucose, see Chap. 4.

1.2 Towards the Operational Formalism in Biological Systems

By understanding that the functional relation (1.1) is either impossible to construct or too complicated to be useful even having been constructed, researchers can try to develop a symbolic calculus of observables (without attempting to relate them to the intrinsic variables). And it seems that sooner or later they will arrive to the formalism in quantum mechanics—the formal calculus of observables.[3]

We recall that W. Heisenberg proceeded precisely in this way, see Chap. 1 of monograph [1] for the historical review. At the very beginning of the development of quantum theory many attempts were made to embed quantum observations into the classical physical framework, e.g., *Bohr-Kramers-Slater theory.* However, all such models were unsatisfactory. Then Heisenberg decided to forget about functional relations of the type (1.1) and to develop a formal calculus of observables. He represented observables by matrices. Since in general matrices do not commute, his operational formalism for measurable quantities was based on noncommutative calculus. In the abstract framework (which was established later by Dirac [2] and Von Neumann [3]) observables are represented by operators. Mathematical details can be found in the book [4].

The quantum formalism cannot say anything about the behavior of a concrete quantum system. For example, nobody is able to predict when the electron in the excited atom will emit the photon and fall to the ground state. This formalism predicts probabilities for the results of measurements. In particular, it predicts the probability of the aforementioned event in an ensemble of atoms in the excited state.

In principle, one may dream (as A. Einstein did) of a theory of the classical type operating with intrinsic variables of quantum systems (which are not just symbolic operator quantities, but take definite values) and representing measurable quantities in the functional form (1.1). A possibility of creation of such "beyond quantum" models is still a subject of debates.

Thus bio-scientists (cellular biologists, evolutionary biologists, cognitive scientists, brain researchers, psychologists) have to try to use the quantum formalism to provide the operational description of bio-observables. By proceeding in this way they will automatically accept the usage of *quantum information* theory in biological science.

Of course, operational formalisms for measurable quantities are not restricted to the quantum one. Even in quantum physics the original (Dirac-von Neumann) model based on Hermitian operators in a complex Hilbert space was generalized in various ways. We can mention the usage of *positive operator valued measures* to represent observables (instead of Hermitian operators). Nowadays this is the standard tool of quantum information theory, see Chap. 6 for applications to cognitive science and decision making. Moreover, one can guess that the known generalizations of the

---

[3] Originally this formalism was developed for observables in microsystems. The main point of this book is that the formalism is not rigidly coupled to microsystems. This is a very general formalism describing observables. And it can be applied to observables of any kind.

quantum formalism are not sufficient to model all possible biological phenomena. And we shall show that this is really the case. In this book (Chap. 4) we present a novel extension of the quantum formalism based on such an advanced tool of quantum information theory as *lifting*.

1.3 Contextuality of Quantum Physics and Biology

According to N. Bohr, quantum physics is fundamentally contextual. Quantum observables cannot be assigned to quantum systems. Results of measurements are combinations of inputs from quantum systems and measurement devices. For example, observables of position and momentum cannot be assigned to a quantum system alone. The statement "the quantum system has the position x" has the meaning that by using a measurement device suitable for position measurement we have obtained this result. Roughly speaking, before measurement a quantum system "did not know" its position and only in the process of interaction with the measurement device this value of "position" was created. Thus quantum observables are not objective, their values cannot be assigned to quantum systems before measurement. By considering the value of, e.g., position one has to specify the *context of measurement* determining this value. N. Bohr emphasized that the whole experimental arrangement has to be taken into account.

This viewpoint on quantum observables matches very well with the situation in biology and, especially, in cognitive science, psychology, decision making. In cognitive science and psychology nonobjectivity of at least some "mental observables" is unquestionable. We need not appeal to an advanced theory to understand that by asking a question to a person, say Alice (a very popular personage in illustrative examples of quantum information theory) we cannot assume that Alice had the definite answer to this question before she has been asked. The answer (the result of measurement) is created in the processes of information processing in Alice's brain and it depends crucially on the measurement context[4]—in particular, on the way, in which the question has been formulated. The latter is the well known *framing effect* which has attracted a lot of attention in cognitive psychology, see [5, 6] for reviews and coupling with the QL-approach. Hence, the contextuality of quantum mechanics is its important commonality with biological science. This is also an important argument in favor of using the quantum formalism in biology. Contextuality is automatically encoded in this formalism.

We point to one important difference between quantum physics and quantum bio-informatics. In the former, observer is always external with respect to the quantum systems under observation.[5] However, bio-systems are able to perform

[4] This question was discussed in detail in *Gestalt psychology*.

[5] In particular, this implies severe problems in attempts to create quantum cosmology. Who is an external observable making measurement of the universe? This problem is the root of a few interpretations which differ from the operational interpretation of quantum physics; for example, some exotic interpretations such as the *many worlds interpretation*.

self-measurements. For example, the brain need not get questions and problems from the outside world to start preparing answers. It can ask itself and answer to itself. Hence, in quantum bio-informatics we cannot avoid the consideration of *self-measurements*. This is a real mismatch with the ideology of quantum physics (at least with the operational interpretation). This foundational problem of quantum bio-informatics has to be studied in more detail. In this book we proceed very pragmatically. We stress the functional complexity of human brain. Therefore it is natural to suppose that one functional unit can "ask questions and get answers" from other functional units. (Of course, this viewpoint needs more justification from neurophysiology.) Thus, opposite to quantum physics, in quantum bio-informatics self-measurements are acceptable.

The aforementioned arguments work well for cognitive systems. For cellular biology, the idea of self-measurement implies a fundamental interpretational problem. Opposite to biological organisms of higher level, the problem of *cell cognition* has not been studied so much, cf. Karafyllidis [7]. It seems that a cell has some form of cognition, it seems that it can "ask questions" to itself (e.g., about its own state or the states of other cells) and "get answers". However, it is not clear how far we can proceed with the analogy between the cognition of, e.g., animals and "cognition" of cells. (In principle, we can define cell cognition as the ability to perform aforementioned self-measurements.)

Finally, we discuss the contextuality in cellular biology. Cell behavior is evidently contextual. For example, consider *cell differentiation*. This process depends crucially on cellular context, especially in the form of signaling from other cells. The same situation we have for genes expressions; the level of the expression of a special gene cannot be considered outside the context of expressions of other genes.

Does the contextuality of bio-observables imply that bio-systems do not have objective properties at all? The answer is definitely negative! Bio-systems have objective properties, but QL states (encoded via pure states or in general density operators) do not specify these properties. Suppose that the levels of gene expressions in an ensemble of cells are represented as the pure state ψ. This state does not describe the real situation in a single cell. It only describes potential levels of genes expressions in the ensemble of cells. It describes predictions for coming measurements. One may say that each single cell "knows" its levels of genes expressions (for this cell these are objective quantities). However, a geneticist does not know and ψ represents uncertainty in the geneticist's knowledge about genes' expressions in this ensemble of cells.

This is appropriate to note that on many occasions Niels Bohr emphasized that quantum mechanics is not about physical processes in microworld, but about our measurements [8]:

> Strictly speaking, the mathematical formalism of quantum mechanics and electrodynamics merely offers rules of calculation for the deduction of expectations pertaining to observations obtained under well-defined experimental conditions specified by classical physical concepts.

In the same way, quantum bio-informatics is not about biological processes in bio-systems, but about results of possible measurements on these systems.

1.4 Adaptive Dynamical Systems

Bio-systems are fundamentally adaptive. They could not survive without developing adaptive skills. They live in the permanently changing environment and the adaptivity to new contexts is really the question of survival. Therefore it is natural to model, e.g., the process of evolution by adaptive dynamical systems, Chap. 8. In the same way processing of information by the brain is described by the adaptive dynamics. The brain permanently updates its dynamical system by taking into account variability of mental contexts. Roughly speaking, by asking a question to Alice, we immediately change the dynamical system operating in Alice's brain. The question is processed by a new dynamical system which takes into account a new context, the context of this question. (See Chap. 6 for the corresponding mathematical model of decision making in the games of the Prisoners Dilemma type.)

In physics the mathematical formalization of the adaptive dynamics (AD) has implicitly appeared in series of papers [4, 9–17] for the study of compound quantum dynamics, chaos, and the quantum realization of the algorithm on the satisfiability problem (*SAT algorithm*). The name of the adaptive dynamics was deliberately used in [17]. The AD has two aspects, one of which is the "observable-adaptive" and another is the "state-adaptive". We now present very general statements about these two types of adaptivity. At the moment these statements can make the impression of cryptograms. The precise contents of these cryptograms will become evident from their mathematical representation in Chap. 4.

The observable-adaptive dynamics is a dynamics characterized as follows:

(1) *Measurement depends on how to see an observable to be measured.*
(2) *The interaction between two systems depends on how a fixed observable exists, that is, the interaction is related to some aspects of observables to be measured or prepared.*

The state-adaptive dynamics is a dynamics characterized as follows:

(1) *Measurement depends on how the state and observable to be used exist.*
(2) *The correlation between two systems depends on how the state of at least one of the systems exists, e.g., the interaction Hamiltonian depends on the state.*

The idea of observable-adaptivity comes from studying chaos. We have claimed that any observation will be unrelated or even contradictory to mathematical universalities such as taking limits, sup, inf, etc. The observation of chaos is a result due to taking suitable scales of, for example, time, distance or domain, and it will not be possible in limiting cases. Examples of the observable-adaptivity are used to understand chaos [10, 15] and examine the violation of Bell's inequality, namely, the chameleon dynamics of Accardi [18]. The idea of the state-adaptivity is implicitly used in the construction of a compound state for quantum communication [9, 11, 19, 20]. Examples of the state-adaptivity can be seen in an algorithm solving NP complete problem, i.e., a pending problem for more than 30 years asking whether there exists an algorithm solving a NP complete problem in polynomial time, as discussed [4, 13, 16].

The mathematical details of the adaptive dynamics will be discussed in Chap. 4.

1.5 Breaking the Formula of Total Probability and Non-Kolmogorov Probability Theory

Probability theory founded by Andrei N. Kolmogorov is a very powerful tool, which can be applicable to various fields. Especially the *formula of total probability* (FTP) plays the crucial role in the *Bayesian approach* to decision making. Nowadays this approach dominates in probabilistic mathematical models of cognitive science, psychology, sociology, cellular biology, genetics.

We present the essence of the usage of FTP[6] in decision making. Suppose that a biosystem is able to estimate probabilities of occurrences of the events A_i $(i = 1, 2, \ldots)$, and $P(A_i)$ are given as subjective probabilities or as empirical probabilities (e.g., frequencies). Suppose also that this bio-system can estimate conditional probabilities: $P(B_j|A_i)$, i.e., the probability that the event B_j occurs under the condition that A_i takes place. Then this bio-system estimates the total probability $P(B_j)$ by using the FTP:

$$P\left(B_j\right) \equiv P\left(B_j | \cup_i A_i\right) = \sum_i P\left(A_i\right) P\left(B_j | A_i\right), \tag{1.2}$$

The interrelation between the magnitudes of probabilities $P(B_j)$ determines the decision strategy. In the simplest case $j = 1, 2$, the bio-system has to select between these two strategies, either B_1 or B_2. This system splits the total (unconditional) probabilities $P(B_1)$ and $P(B_2)$ by using FTP and if, e.g., $P(B_1)$ is larger than $P(B_2)$, then the bio-system selects the strategy B_1. (If $P(B_1) \approx P(B_2)$, then this approach would not work, additional information has to be collected.) Typically "elementary events" A_i are sufficiently simple. To find the conditional probabilities $P(B_j|A_i)$ is easier than the total probability $P(B_\beta)$.

However, in general, there is no reason to assume that all bio-systems and in all situations act by using this Bayesian scheme. The simplest argument against the usage of (1.2) is that in some situations a bio-system does not have enough computational resources to proceed in the Bayesian way. A decision strategy has to be selected very quickly: "to decide or to die". A bio-system does not have time first to estimate

---

[6] The Kolmogorov probability model based on the representation of events by sets and probabilities by measures will be presented in Chap. 2. For the moment, we operate with probabilities at the formal level. We remark that in the rigorous framework FTP is a theorem.

probabilities of elementary conditions and corresponding conditional probabilities and then sum up. It makes an "integral decision" by using other decision schemes. The integral probability $P(B_j)$ can differ from the one obtained through decomposition into elementary conditions. Hence, in some situations FTP can be violated, although, from the formal mathematical viewpoint based on the Kolmogorov model, this seems to be completely impossible.

The latter implies that it is impossible to define the integral probability and decomposition probabilities in a single Kolmogorov space. See Chap. 2 for the formal presentation of the notion of Kolmogorov space. Here we just remark that it is a collection of sets representing events and a probability measure defined on it. The corresponding statistical data have to be described by a *non-Kolmogorovian model*. Typically a non-Kolmogorovian model can be represented as a family of Kolmogorov probability spaces coupled in some way—to provide a possibility to express probabilities of one space through probabilities of another space. S. Gudder treated such structures as *probability manifolds*. We can say that the violation of Eq. (1.2) implies the impossibility to find a single Kolmogorov space, in which both random variables for events A_i and events B_j are well defined. In [21] such random variables were called *probabilistically incompatible*.

In bioscience we have plenty of statistical data violating FTP: recognition of ambiguous figures [22, 23], disjunction effect in cognitive psychology [24–34], interference of genes' expressions [35–37]. It seems that in decision making bio-systems are fundamentally non-Bayesian. Hence, in general, we cannot proceed with FTP and the Kolmogorov model. Other probabilistic formalisms and decision making schemes have to be tested to match with biological behavior.

In the quantum probabilistic model (where probabilities are defined via diagonal elements of density matrices, see (2.28)) FTP is violated[7] [21]; the model is non-Kolmogorovian. This is another reason to use the quantum operational formalism in biology. (Empirical data tell us: we cannot proceed with classical (Kolmogorov) probabilities. Therefore we have to test other probabilistic models, and quantum probability is most well-known).

In quantum physics we cannot measure all observables jointly. The impossibility of joint measurement is operationally represented as the noncommutativity of operators representing observables. In fact, quantum incompatible observables are probabilistically incompatible [21]. They cannot be defined for a single Kolmogorov probability space. Thus quantum probability theory is a non-Kolmogorovian model. It is represented as a collection of Kolmogorov probability spaces. We will discuss the Kolmogorov probability theory in Chap. 2, see also Chap. 4.

We summarize this rather long discussion on non-Kolmogorovness of quantum probability (and, hence, non-Bayesian structure of the corresponding theory of decision making).

---

[7] The basic quantum experiment demonstrating the violation of FTP is the *two slit experiment* demonstrating interference of probabilistic patterns related to two incompatible experimental contexts: one of slits is open and the other is closed, see Sect. 4.1.1. In general, violation of FTP means the presence of nontrivial interference of probabilistic patters.

Quantum probability is based on the usage of a family of Kolmogorov probability spaces corresponding to incompatible observables. Although it is impossible to perform joint measurements of such observables, the theoretical model provides a possibility to couple probabilistic data collected for them.

1.6 Quantum Bio-informatics

We motivate applications of the formalism of quantum mechanics and, in particular, quantum information to biology (in the broad sense, including cognitive science and psychology) without reductionism of biological processes to quantum physical processes in bio-systems (from cells to brains). Our motivation is based on the will to proceed by ignoring details and using an operational (symbolic) description of statistical data collected in measurements. Thus this book is not about *quantum biology*; we shall not study quantum physical processes in cells or try to model brain functioning from quantum physics of the brain as a material body, cf. Penrose [5, 6], Hameroff [38, 39], and [40]. To distinguish our modeling from real quantum physics (and attempts to apply it to biology), we use the terminology *quantum-like* (QL), instead of simply quantum, more generally, adaptive dynamics. To emphasize the biological dimension of our QL-modeling an adaptive dynamics, we shall often use the terminology *quantum-like bio-informatics* or simply *quantum bio-informatics*. In general, quantum bio-informatics does not deny attempts to reduce information processing in bio-systems to quantum physical information processing, i.e., the one based on quantum physical systems as carriers of information. The latter is an exciting project; and there is still no reason to reject it completely (although this project has been criticized a lot, mainly for the incompatibility of spatial, temporal, and thermodynamical scales of quantum physical and biological processes, see, e.g., [41]). However, in general, quantum bio-informatics has no direct relation to quantum biology (including models of quantum brain[8]).

In this book we shall show that the biological phenomena including adaptation (*E. coli* glucose/lactose growth), differentiation/embryogenesis (e.g., tooth regeneration), scrapie (prion aggregation), cognition (long/short judgment of bars with or without preflashing circles), epimutation and usual mutations, and even including ecological systems can be described similarly by the quantum bio-informatics approach, i.e., adaptive dynamical lifting.

---

[8] Cf. with QL-models of brain functioning [1, 42–47].

References

1. Haven, E., Khrennikov, A.: Quantum Social Science. Cambridge Press, Cambridge (2012)
2. Dirac, P.A.M.: The Principles of Quantum Mechanics. Clarendon Press, Oxford (1995)
3. Von Neuman, J.: Mathematical Foundations of Quantum Mechanics. Princeton University Press, Princeton (1955)
4. Ohya, M., Volovich, I.V.: Mathematical Foundations of Quantum Information and Computation and Its Applications to Nano- and Bio-systems. Springer, Berlin (2011)
5. Khrennikova, P., Khrennikov, A., Haven, E.: The quantum-like description of the dynamics of party governance in the US political system (2012). arXiv:1206.2888 [physics.gen-ph]
6. Khrennikova, P.: Evolution of quantum-like modeling in decision making processes. AIP Conf. Proc. **1508**, 108–112 (2012)
7. Karafyllidis, I.: Quantum mechanical model for information transfer from DNA to protein. Biosystems **93**, 191–198 (2008)
8. Bohr, N.: The Philosophical Writings of Niels Bohr, vol. 3. Ox Bow Press, Woodbridge (1987)
9. Ohya, M.: On compound state and mutual information in quantum information theory. IEEE Trans. Inf. Theory **29**, 770–777 (1983)
10. Ohya, M.: Complexities and their applications to characterization of chaos. Int. J. Theor. Phys. **37**, 495–505 (1998)
11. Accardi, L., Ohya, M.: Compound channels, transition expectations, and liftings. Appl. Math. Optim. **39**, 33–59 (1999)
12. Inoue, K., Ohya, M., Sato, K.: Application of chaos degree to some dynamical systems. Chaos, Soliton and Fractals **11**, 1377–1385 (2000)
13. Ohya, M., Volovich, I.V.: New quantum algorithm for studying NP-complete problems. Rep. Math. Phys. **52**, 25–33 (2003)
14. Inoue, K., Ohya, M., Volovich, I.V.: Semiclassical properties and chaos degree for the quantum baker's map. J. Math. Phys. **43** (2002)
15. Kossakowski, A., Ohya, M., Togawa, Y.: How can we observe and describe chaos? Open Syst. Inf. Dyn. **10**(3), 221–233 (2003)
16. Accardi, L., Ohya, M.: A Stochastic limit approach to the SAT problem. Open Syst. Inf. Dyn. **11**, 1–16 (2004)
17. Ohya, M.: Adaptive dynamics and its applications to chaos and NPC problem. QP-PQ: quantum probability and white noise analysis. Quantum Bio-Inform. **21** 181–216 (2007)
18. Accardi, L.: Urne e camaleonti: Dialogo sulla realta, le leggi del caso e la teoria quantistica.Il Saggiatore, Rome (1997); English edn. World Scientific, Singapore (2002); Japanese edn, Makino, Tokyo (2002); Russian edn. Regular and Chaotic dynamics, Ijevsk (2002)
19. Ohya, M.: Note on quantum probability. Il Nuovo Cimento **38**, 203–206 (1983)
20. Ohya, M.: Some aspects of quantum information theory and their applications to irreversible processes. Rep. Math. Phys. **27**, 19–47 (1989)
21. Khrennikov, A.: Contextual Approach to Quantum Formalism. Springer, Berlin (2009)
22. Conte, E., Todarello, O., Federici, A., Vitiello, F., Lopane, M.: A preliminary evidence of quantum-like behaviour in measurements of mental states. Quantum Theory: Reconsideration of Foundations. Series Mathematical Modeling Physics, Engineering, and Cognitive Science **10**, 679–702 (2004)
23. Conte, E., Khrennikov, A., Todarello, O., Federici, A., Zbilut, J.P.: Mental states follow quantum mechanics during perception and cognition of ambiguous figures. Open Syst. Inf. Dyn. **16**, 1–17 (2009)
24. Shafir, E., Tversky, A.: Thinking through uncertainty: nonconsequential reasoning and choice. Cogn. Psychol. **24**, 449–474 (1992)
25. Tversky, A., Shafir, E.: The disjunction effect in choice under uncertainty. Psychol. Sci. **3**, 305–309 (1992)
26. Hofstader, D.R.: Dilemmas for superrational thinkers, leading up to a luring lottery. Sci. Am. **6** (1983)

27. Hofstader, D.R.: Metamagical Themes: Questing for the Essence of Mind and Pattern. Basic Books, New York (1985)
28. Croson, R.: The disjunction effect and reasoning-based choice in games. Organ. Behav. Hum. Decis. Process. **80**, 118–133 (1999)
29. Busemeyer, J.B., Wang, Z., Townsend, J.T.: Quantum dynamics of human decision making. J. Math. Psychol. **50**, 220–241 (2006)
30. Busemeyer, J.R., Matthews, M., Wang, Z.: A quantum information processing explanation of disjunction effects. In: Sun, R., Myake, N. (eds.) The 29th Annual Conference of the Cognitive Science Society and the 5th International Conference of Cognitive Science, pp. 131–135. Mahwah (2006)
31. Busemeyer, J.R., Santuy, E., Lambert-Mogiliansky, A.: Comparison of Markov and quantum models of decision making. In: Bruza, P., Lawless, W., van Rijsbergen, K., Sofge, D.A., Coeke, B., Clark, S. (eds.) Quantum Interaction: Proceedings of the Second Quantum Interaction Symposium, pp. 68–74. College Publications, London (2008)
32. Busemeyer, J., Bruza, P.D.: Quantum Cognition and Decision. Cambridge University Press, Cambridge (2012)
33. Pothos, E.M., Busemeyer, J.R.: A quantum probability explanation for violation of rational decision theory. Proc. R. Soc. **276**(1165), 2171–2178 (2009)
34. Wang, Z., Busemeyer, J.R.: A quantum question order model supported by empirical tests of an a priori and precise prediction. Top. Cogn. Sci. (2013), to be published
35. Asano, M., Basieva, I., Khrennikov, A., Ohya, M., Tanaka, Y., Yamato, I.: Quantum-like model for the adaptive dynamics of the genetic regulation of *E. coli*'s metabolism of glucose/lactose. Syst. Synth. Biol. **6**, 1–7 (2012)
36. Asano, M., Basieva, I., Khrennikov, A., Ohya, M., Tanaka, Y., Yamato, I.: Quantum-like model of diauxie in *Escherichia coli*: operational description of precultivation effect. J. Theor. Biol. **314**, 130–137 (2012)
37. Asano, M., Basieva, I., Khrennikov, A., Ohya, M., Yamato, I.: Non-Kolmogorovian approach to the context-dependent systems breaking the classical probability law. Found. Phys. **43**(7), 895–911 (2013)
38. Hameroff, S.: Quantum coherence in microtubules. A neural basis for emergent consciousness? J. Conscious. Stud. **1**, 91–118 (1994)
39. Hameroff, S.: Quantum computing in brain microtubules? The Penrose-Hameroff Orch Or model of consciousness. Philos. Trans. R. Soc. Lond. A 1–28 (1994)
40. Fichtner, K.-H., Fichtner, L., Freudenberg, W., Ohya, M.: On a quantum model of the recognition process. QP-PQ: Quantum Prob. White Noise Anal. **21**, 64–84 (2008)
41. Donald, M. J.: A Review of Quantum Evolution (2001). arXiv:quant-ph/0101019
42. Khrennikov, A.: Quantum-like brain: interference of minds. BioSystem **84**, 225–241 (2006)
43. Khrennikov, A.: The quantum-like brain on the cognitive and subcognitive time scales. J. Conscious. Stud. **15**, 39–77 (2008)
44. Khrennikov, A.: On quantum-like probabilistic structure of mental information. Open Syst. Inf. Dyn. **11**, 267–275 (2004)
45. Khrennikov, A.: Ubiquitous Quantum Structure: From Psychology to Finance. Springer, Heidelberg (2010)
46. Khrennikov, A.: On the physical basis of theory of "mental waves". Neuroquantology **8**, S71–S80 (2010)
47. Khrennikov, A.: Quantum-like model of processing of information in the brain based on classical electromagnetic field. Biosystem **105**(3), 250–262 (2011)

Chapter 2
Fundamentals of Classical Probability and Quantum Probability Theory

Abstract In this chapter we present briefly the basic notions of classical and quantum theories of probability and information. This chapter is especially important for biologists, psychologists, experts in cognition, and sociologists who were not trained in quantum theory (but even classical theory is presented in a simple manner). We start with the presentation of the standard measure-theoretic formulation of the modern classical probability theory (Kolmogorov, Grundbegriffe der Wahrscheinlichkeitsrechnung. Springer, Berlin [1]). Then we turn to fundamentals of quantum formalism, including theory of open quantum systems and its generalizations.

Keywords Foundations of classical and quantum probability · Quantum formalism · Quantum states and observables · Quantum information · Born rule for probabilities · Quantum master equation

2.1 Short Introduction to Classical Probability Theory

2.1.1 Probability Space

The modern axiomatics of probability theory was invented by Andrei Nikolaevich Kolmogorov in 1933 [1]. A crucial point is the *representation of events by subsets* of some basic set Ω. The collection of subsets representing events should be sufficiently rich—to be able to perform set-theoretic operations such as the intersection, the union, and the difference of sets. However, at the same time it should be reasonably rich. If a too extended system of subsets is selected to represent events, then it may contain "events" which cannot be interpreted in a reasonable way. After selection of a proper system of sets to represent events, one assigns weights to these subsets:

$$A \mapsto P(A). \tag{2.1}$$

The probabilistic weights are chosen to be nonnegative real numbers and normalized by 1: $P(\Omega) = 1$, the probability that something happens equals one. An event with large weight is more probable than an event with small weight. The weight of an

M. Asano et al., *Quantum Adaptivity in Biology: From Genetics to Cognition*,
DOI 10.1007/978-94-017-9819-8_2

event A that can be represented as the disjoint union of events A_1 and A_2 is equal to the sum of weights of these events. The latter property is called *additivity*. (There is the evident similarity with mass, area, volume.)

It is useful to impose some restrictions on the system of sets representing events: (a) set Ω containing all possible events and the empty set $\emptyset$ are events (something happens and nothing happens); (b) the union of two sets representing events represents an event; (c) the intersection of two sets representing events represents an event; (d) the complement of a set representing an event, i.e., the collection of all points that do not belong to this set, again represents an event. These set-theoretic operations correspond to the basic operations of (Boolean) logic: "or", "and", "no". And the modern set-theoretic representation of events is a mapping of propositions describing events onto sets with preservation of the logical structure. At the beginning of the mathematical formalization of probability theory the map (2.1) was defined on an algebraic structure corresponding to the logical structure, the *Boolean algebra* (invented by Boole, the creator of "Boolean logic" [2]). The set-system with properties (a)–(d) is called the *algebra of sets* (in the American literature, the field of sets).

In the case of finite Ω the map given by (2.1) with the above-mentioned properties is (measure-theoretic) probability. (Since Ω can contain billions of points, this model is useful in a huge class of applications.) Here $\Omega = \{\omega_1, \ldots, \omega_N\}$. To determine any map (2.1), it is enough to assign to each elementary event its weight

$$0 \leq P(\omega_j) \leq 1, \ \sum_j P(\omega_j) = 1.$$

Then by additivity this map is extended to the set-algebra consisting of all subsets of Ω:

$$P(A) = \sum_{\{\omega_j \in A\}} P(\omega_j).$$

However, if Ω is countable, i.e., it is infinite and its points can be enumerated, or "continuous"—e.g., a segment of the real line $\mathbf{R}$, then simple additivity is not sufficient to create a fruitful mathematical model. The map (2.1) has to be additive with respect to countable unions of disjoint events:

$$P(A_1 \cup \ldots \cup A_n \cup \ldots) = P(A_1) + \cdots + P(A_n) + \cdots, \tag{2.2}$$

and to work fruitfully with such maps (e.g., to integrate), one has to impose special restrictions on the system of sets representing events. It has to be not simply a set-algebra, but a σ-algebra of sets (in the American literature, a σ-field), i.e., (b) and (c) must be valid for countable unions and intersections of sets. In logical terms, it means that the operations "or" and "and" can be applied infinitely many times to form new events. Of course, this is a mathematical idealization of the real situation. One of the most important "continuous probability model" is based on $\Omega = \mathbf{R}$, i.e., elementary events are represented by real numbers. Typically a σ-algebra is selected as the *Borel σ-algebra:* it is generated by all half-open intervals, $[\alpha, \beta), \alpha < \beta$, with the aid of the operations of the union, intersection, and complement.

We remark that although probability models with infinite spaces of elementary events play an important role in many applications, one can, in principle, read practically the whole book using finite Ω, the collection of events represented by all its subsets, and additive probability given by assigning weights to points of Ω.

Let Ω be a set and let $\mathcal{F}$ be a σ-algebra of its subsets. A *probability measure* **P** is a map from $\mathcal{F}$ to the segment $[0, 1]$ normalized $P(\Omega) = 1$ and σ-additive, i.e., the equality (2.2) holds for disjoint sets belonging to $\mathcal{F}$.

By the Kolmogorov axiomatics [1] the *probability space* is a triple

$$\mathcal{P} = (\Omega, \mathcal{F}, P).$$

Points ω of Ω are said to be *elementary events*, elements of $\mathcal{F}$ are *events*, P is *probability*.

Random observations are represented by random variables. We start with mathematically simplest random variables, the discrete ones. In fact, the majority of observables (classical and quantum) considered in this book are discrete-valued.

Discrete random variables on the Kolmogorov space $\mathcal{P}$ are by definition functions $a : \Omega \to X_a$, where $X_a = \{\alpha_1, \ldots, \alpha_n, \ldots\}$ is a countable set (the *range of values*) such that the sets

$$C^a_\alpha = \{\omega \in \Omega : a(\omega) = \alpha\}, \alpha \in X_a, \tag{2.3}$$

belong to $\mathcal{F}$.

It is typically assumed that the range of values X_a is a subset of the real line. We will proceed under this assumption practically everywhere, but sometimes, e.g., in cognitive and psychological modeling, it will be more convenient to consider Boolean labels, e.g., $\alpha =$ yes, no.

The probability distribution of a (discrete) random variable a is defined as

$$p^a(\alpha) \equiv P(\omega \in \Omega : a(\omega) = \alpha).$$

We remark that

$$p^a(\alpha_1) + \cdots + p^a(\alpha_n) + \cdots = 1, \;\; p^a(\alpha_n) \geq 0. \tag{2.4}$$

The *average* (mathematical expectation) of a random variable a is defined as

$$\bar{a} \equiv Ea = \alpha_1 \; p^a(\alpha_1) + \cdots + \alpha_n \; p^a(\alpha_n) + \cdots . \tag{2.5}$$

If the set of values of ξ is infinite, then the average is well defined if the series in the right-hand side of (2.5) converges absolutely.

For a family of random variables $a_1, \ldots, a_m$, their *joint probability distribution* is defined as

$$p^{a_1 \cdots a_m}(\alpha^1_{j_1}, \ldots, \alpha^m_{j_m}) = P(\omega \in \Omega : a_1(\omega) = \alpha^1_{j_1}, \ldots, a_m(\omega) = \alpha^m_{j_m}). \tag{2.6}$$

We remark that the joint probability is symmetric with respect to permutations; e.g., for two random variables a and b, we have

$$p^{ab}(\alpha, \beta) = P(\omega \in \Omega \colon a(\omega) = \alpha, b(\omega) = \beta) = p^{ba}(\beta, \alpha). \tag{2.7}$$

For two (discrete) random variables a and b, their *covariance* is defined as

$$\mathrm{cov}(a, b) = E(a - \bar{a})(b - \bar{b}) = \sum_{\alpha\beta} (\alpha - \bar{a})(\beta - \bar{b})\; p^{ab}(\alpha, \beta). \tag{2.8}$$

It is easy to see that

$$\mathrm{cov}(a, b) = \overline{ab} - \bar{a}\bar{b}. \tag{2.9}$$

We remark that covariance is symmetric

$$\mathrm{cov}(a, b) = \mathrm{cov}(b, a). \tag{2.10}$$

Suppose now that the results of random measurement cannot be represented by a finite or countable set. Thus such an observable cannot be represented as a discrete random variable.

A *random variable* on the Kolmogorov space $\mathscr{P}$ is by definition any function $\xi : \Omega \to \mathbf{R}$ such that for any set Γ belonging to the Borel σ-algebra, its pre-image belongs to the σ-algebra of events $\mathscr{F} : \xi^{-1}(\Gamma) \in \mathscr{F}$, where $\xi^{-1}(\Gamma) = \{\omega \in \Omega : \xi(\omega) \in \Gamma\}$.

In this case the mathematical formalism is essentially more complicated. The main mathematical difficulty is to define the integral with respect to a probability measure on $\mathscr{F}$, the *Lebesgue integral* [1]. In fact, classical probability theory, being based on measure theory, is mathematically more complicated than quantum probability theory. In the latter integration is replaced by the trace operation for linear operators and the trace is always the discrete sum. Here we are not able to present theory of Lebesgue integration. We formally invent the symbol of integration.[1]

[1] We remark that, for a discrete random variable, the integral coincides with the sum for the mathematical expectation, see (2.5). And a discrete random variable is integrable if its mathematical expectation is well defined. In general any integrable random variable can be approximated by integrable discrete random variables and its integral is defined as the limit of the integrals for the approximating sequence.

The *average* (mathematical expectation) of a random variable a is defined as

$$\bar{a} \equiv Ea = \int_{\Omega} \xi(\omega) dP(\omega). \tag{2.11}$$

The probability distribution of a random variable a is defined (for Borel subsets of the real line) as

$$p^a(\Gamma) \equiv P(\omega \in \Omega : a(\omega) \in \Gamma).$$

This is a probability measure on the Borel σ-algebra. And the calculation of the average can be reduced to integration with respect to p^a:

$$\bar{a} \equiv Ea = \int_{\mathbf{R}} x \, dp^a(x). \tag{2.12}$$

For simplicity, further we shall proceed only with discrete random variables. The general case can be treated by using everywhere integrals instead of sums.

2.1.2 Conditional Probability

Kolmogorov's probability model is based on a probability space equipped with the operation of conditioning. In this model *conditional probability* is defined by the well known *Bayes' formula*

$$P(B|C) = P(B \cap C)/P(C), \quad P(C) > 0. \tag{2.13}$$

By Kolmogorov's interpretation it is the *probability of an event B to occur under the condition that an event C has occurred.*

We remark that $P_C(B) \equiv P(B|C)$ is again a probability measure on $\mathscr{F}$. For a set $C \in \mathscr{F}$, $P(C) > 0$, and a (discrete) random variable a, the conditional probability distribution is defined as

$$p_C^a(\alpha) \equiv P(a = \alpha|C).$$

We naturally have

$$p_C^a(\alpha_1) + \cdots + p_C^a(\alpha_n) + \cdots = 1, \; p_C^a(\alpha_n) \geq 0. \tag{2.14}$$

The conditional expectation of a random variable a is defined by

$$E(a|C) = \alpha_1 \; p_C^a(\alpha_1) + \cdots + \alpha_n \; p_C^a(\alpha_n) + \cdots \tag{2.15}$$

For two random variables a and b, consider conditional probabilities

$$p_{\alpha\beta} \equiv P(b = \beta|a = \alpha), \; p_{\beta\alpha} \equiv P(a = \alpha|b = \beta).$$

These conditional probabilities are called *transition probabilities*. (This terminology is also used in quantum probability, Sect. 4.1.1.)

These conditional probabilities can also be written in the form

$$p_{\alpha\beta} = P(b = \beta | C^a_\alpha), \ p_{\beta\alpha} = P(a = \alpha | C^b_\beta), \tag{2.16}$$

where, e.g., C^a_α is defined by (2.3). It is, of course, assumed that in the first case $p^a(\alpha) > 0$ and in the second case $p^b(\beta) > 0$.

It is useful to consider the matrices of transition probabilities

$$\mathbf{P}^{b|a} = (p_{\alpha\beta}), \ \mathbf{P}^{a|b} = (p_{\beta\alpha}). \tag{2.17}$$

We remark that these matrices are always *left stochastic*. The left stochastic matrix is a square matrix whose columns consist of nonnegative real numbers whose sum is 1. For example, for the matrix $\mathbf{P}^{b|a}$, we have

$$\sum_\beta p_{\alpha\beta} = \sum_\beta P(b = \beta | a = \alpha) = 1. \tag{2.18}$$

for any fixed $a = \alpha$. This is a consequence of the fact that, for any set C of strictly positive probability, P_C is also a probability measure.

We point to the following equality connecting the joint probability distribution of two random variables a and b with their transition probabilities

$$p^{ab}(\alpha, \beta) = p^a(\alpha) p_{\alpha\beta} = p^b(\beta) p_{\beta\alpha} = p^{ba}(\beta, \alpha). \tag{2.19}$$

In our further considerations one special class of matrices of transition probabilities will play an important role. These are so called doubly stochastic matrices. We recall that in a *doubly stochastic matrix* all entries are nonnegative and in all rows and all columns they sum to 1. Of course, in general, $\mathbf{P}^{b|a}$ is only left stochastic, but not doubly stochastic. In Sect. 4.1.1 we shall see that for quantum probability, the matrices of transition probabilities are always doubly stochastic for observables of the most important for applications class, observables with non-degenerate spectra.

Consider now a pair of dichotomous random variables $a = \alpha_1, \alpha_2$ and $b = \beta_1, \beta_2$. The matrix of transition probabilities $\mathbf{P}^{b|a}$ has the form

$$\mathbf{P}^{b|a} = \begin{pmatrix} p_{\alpha_1\beta_1} & p_{\alpha_2\beta_1} \\ p_{\alpha_1\beta_2} & p_{\alpha_2\beta_2} \end{pmatrix} \tag{2.20}$$

It is doubly stochastic iff $p_{11} = p_{22}$ and $p_{12} = p_{21}$, i.e.,

$$\mathbf{P}^{b|a} = \begin{pmatrix} p & 1-p \\ 1-p & p \end{pmatrix} \tag{2.21}$$

In particular, it is automatically symmetric.

2.1.3 *Formula of Total Probability*

In our further considerations the important role will be played by the *formula of total probability* (FTP). This is a theorem of the Kolmogorov model. Let us consider a countable family of disjoint sets A_k belonging to $\mathcal{F}$ such that their union is equal to Ω and $P(A_k) > 0, k = 1, \ldots$. Such a family is called a *partition* of the space Ω.

Theorem 2.1 *Let $\{A_k\}$ be a partition. Then, for every set $B \in \mathcal{F}$, the following formula of total probability holds*

$$P(B) = P(A_1)P(B|A_1) + \cdots + P(A_k)P(B|A_k) + \cdots \tag{2.22}$$

Proof We have

$$P(B) = P\big(B \cap \big(\cup_{k=1}^{\infty} A_k\big)\big) = \sum_{k=1}^{\infty} P(B \cap A_k) = \sum_{k=1}^{\infty} P(A_k)\frac{P(B \cap A_k)}{P(A_k)}.$$

Especially interesting for us is the case where a partition is induced by a discrete random variable a taking values $\{\alpha_k\}$. Here,

$$A_k = C^a_{\alpha_k} = \{\omega \in \Omega : a(\omega) = \alpha\}. \tag{2.23}$$

Let b be another random variable. It takes values $\{\beta_j\}$. For any $\beta \in X_b$, we have

$$P(b = \beta) = P(a = \alpha_1)P(b = \beta|a = \alpha_1) + \cdots + P(a = \alpha_k)P(b = \beta|a = \alpha_k) + \cdots \tag{2.24}$$

or in compact notation,

$$p^b(\beta) = p^a(\alpha_1)p_{\alpha_1\beta} + \cdots + p^a(\alpha_k)p_{\alpha_k\beta} + \cdots \tag{2.25}$$

2.2 Short Introduction of Quantum Probability Theory

We shall present the formalism of quantum mechanics (and quantum information) by small portions, from section to section. Advanced notions such as, e.g., channels and lifting, will be introduced in Chap. 4. However, the later notions will be used to construct a generalization of the conventional quantum formalism, which will serve for biological purposes. We start with the presentation of the basic structure of the quantum formalism.

2.2.1 States and Observables

Following Heisenberg, observables are represented by Hermitian matrices or, in the abstract framework, by Hermitian operators. These operators act in the complex Hilbert space $\mathscr{H}$, i.e., a complex linear space endowed with a scalar product denoted as $\langle \psi_1, \psi_2 \rangle$.

Here, let us recall the mathematical definition and properties of scalar product.

The scalar product is a function from the Cartesian product $\mathscr{H} \times \mathscr{H}$ to the field of complex numbers $\mathbb{C}$, $\psi_1, \psi_2 \to \langle \psi_1, \psi_2 \rangle$, having the following properties:

1. Positive definiteness: $\langle \psi, \psi \rangle \geq 0$ with $\langle \psi, \psi \rangle = 0$ if and only if $\psi = 0$.
2. Conjugate symmetry: $\langle \psi_1, \psi_2 \rangle = \overline{\langle \psi_2, \psi_1 \rangle}$[2]
3. Linearity with respect to the first argument:

$$\langle k_1 \psi_1 + k_2 \psi_2, \phi \rangle = k_1 \langle \psi_1, \phi \rangle + k_2 \langle \psi_2, \phi \rangle,$$

where k_1, k_2 are complex numbers.

From the second and third properties it is easy to obtain that for the second argument, $\langle \phi, k_1 \psi_1 + k_2 \psi_2 \rangle = \bar{k}_1 \langle \phi, \psi_1 \rangle + \bar{k}_2 \langle \phi, \psi_2 \rangle$. By fixing in $\mathscr{H}$ an orthonormal basis (e_j), i.e., $\langle e_i, e_j \rangle = \delta_{ij}$, we represent vectors by their coordinates $\psi_1 = (z_1, \ldots, z_n, \ldots)$, $\psi_2 = (w_1, \ldots, w_n, \ldots)$. In the coordinate representation the scalar product has the form $\langle \psi_1, \psi_2 \rangle = \sum_j z_j \bar{w}_j$. By using this representation the reader can easily verify the aforementioned properties of the scalar product.

We remark that the usage of *complex numbers* plays the crucial role. One cannot proceed with real Hilbert spaces. There are experimental statistical data, which cannot be embedded in the real model.

The norm (an abstract analog of the Euclidean length) of a vector is defined as $\|\psi\| = \sqrt{\langle \psi, \psi \rangle}$. In the fixed system of coordinates $\|\psi\| = \sqrt{\sum_j |z_j|^2}$. Normalized vectors of $\mathscr{H}$, i.e., ψ such that $\|\psi\| = 1$, represent a special (and the most important) class of states of quantum systems, *pure states*. Each pure state ψ can be represented as an operator acting in $\mathscr{H}$. which is the orthogonal projections π onto the vector ψ. In terms of the scalar product, the orthogonal projections can be written as $\pi\phi = \langle \phi, \psi \rangle \psi$. By fixing an orthonormal basis in $\mathscr{H}$, the pure state is expressed by a matrix $\rho = (\rho_{ij})$ satisfying

(a) Hermitian: $\rho_{ij} = \bar{\rho}_{ij}$, in particular, the diagonal elements are real,
(b) Positive definiteness: $\langle \rho\phi, \phi \rangle \geq 0$ for any vector ϕ,
(c) Its trace equals 1: $\mathrm{tr}\rho = \sum_j \rho_{jj} = 1$.

The eigenvalues of Hermitian matrix are real numbers. In addition, a Hermitian matrix with the positive definiteness has non-negative eigenvalues. Finally, from the condition c), the sum of all the eigenvalues equals 1.

As an example in the two dimensional space $\mathscr{H} = \mathbb{C}^2$, we introduce the operator given by the following matrix

[2] Here bar denotes complex conjugation; for $z = x + iy$, $\bar{z} = x - iy$.

$$\rho_{\text{qubit}} = \begin{pmatrix} \rho_{11} & \rho_{12} \\ \rho_{21} & \rho_{22} \end{pmatrix} = \begin{pmatrix} |\alpha|^2 & \alpha^*\beta \\ \alpha\beta^* & |\beta|^2 \end{pmatrix} \tag{2.26}$$

for the corresponding pure state vector

$$\psi = \begin{pmatrix} \alpha \\ \beta \end{pmatrix}.$$

Here α and β are complex numbers satisfying $|\alpha|^2 + |\beta|^2 = 1$. The above ρ_{qubit} expresses the state of *quantum bit* (shortly called *qubit*), which is often seen in quantum information theory (see Sect. 2.2.4). One can easily see that ρ_{qubit} satisfies the following condition:

$$\rho_{\text{qubit}}^2 = \rho_{\text{qubit}}. \tag{2.27}$$

This condition guarantees that the operator ρ_{qubit} is a projection on a vector ψ. In general, there are operators that are not projectors but satisfy the conditions (a)–(c).

The next step in the development of the quantum formalism (due to Landau and von Neumann) was proceeding without the Eq. 2.27 constraint, i.e., considering all possible matrices satisfying conditions (a)–(c). They are called *density matrices* and they represent the most general states of quantum systems. In the abstract framework one considers operators satisfying conditions (a)–(c), density operators. Each density operator can be written as a weighted sum of projection operators corresponding to pure states. If such a sum contains more than one element, then the state represented by this density operator is called a *mixed state*: the mixture of pure states with some weights. Although this terminology is widely used, it is ambiguous. The representation of a density operator as a weighted sum of projectors corresponding to pure states is not unique. Thus by using the terminology mixed state one has to take into account this non-uniqueness.

For simplicity let us restrict our consideration to the case of a finite dimensional Hilbert space.

Take an arbitrary quantum observable represented by a Hermitian operator A. For any Hermitian operator, there exists an orthogonal basis of $\mathcal{H}$ consisting of its eigenvectors. The values of the observable operationally represented by A are encoded in the eigenvalues of this operator, $a_1, \ldots, a_n$. (We shall use the same symbol for an observable and its operator representation.)

Consider now the case of nondegenerate spectrum: $a_j \neq a_i$ if $j \neq i$. For an ensemble of quantum systems prepared in the same state represented by the density operator ρ, the probability to obtain a fixed value $P(a_j)$ is encoded in the corresponding matrix element of the operator ρ (in the basis of eigenvectors):

$$P(a_j) = \rho_{jj}. \tag{2.28}$$

This is one of the basic postulates of quantum mechanics. It connects experimental probabilities with the operator representation of observables. (In the simplest form, for the pure state, this postulate was proposed by Born, see (2.32); the form (2.28) is due to von Neumann [3]).

Thus all information about possible results of measurements is encoded in two Hermitian operators, the observable A and the state ρ. This is very compact and convenient representation.

In fact, mathematically, quantum formalism is a linear algebra (with elements of functional analysis for the infinite-dimensional case). Thus it is very simple; there is nothing simpler than a *linear representation.* The corresponding the dynamical equations (e.g., the Schrödinger's equation) are linear.

The simplicity of the quantum linear representation of measurable quantities is one of the reasons to use this formalism, in particular, in biology.

Surprisingly, practically any theory of statistical measurements can be operationally represented in a linear space. Here we have no possibility to discuss this problem that was studied in detail already in 1950th by Mackey; recently one of the authors of this book has proposed a *quantum-like representation algorithm* [4]. This algorithm constructs the QL representation of statistical data collected for two observables (random variables) under some restrictions on the matrix of transition probabilities for these observables. Hence, all "natural operational representations" are reduced to linear representations (in some situations one has to use number systems different from complex numbers, e.g., the so called hyperbolic numbers [4]).

2.2.2 Superposition

Let the state space of some system (physical or biological) be represented as a finite-dimensional Hilbert space $\mathcal{H}$. Consider a pure state ψ and an observable A, denote its eigenvalues by $a_1, \ldots, a_m$ and the corresponding eigenvectors by $e_1, \ldots, e_m$. This is an orthonormal basis in $\mathcal{H}$. (We again proceed under the assumption that all eigenvalues are different.) We expand the vector ψ with respect to this basis:

$$\psi = c_1 e_1 + \cdots + c_m e_m, \tag{2.29}$$

where (c_j) are complex numbers such that the sum of their squared absolute values are equal to one (this is the coordinate expression of the normalization by one of a pure state vector):

$$|c_1|^2 + \cdots + |c_m|^2 = 1. \tag{2.30}$$

By using the terminology of linear algebra we say that the pure state ψ is a *superposition* of pure states e_j.

The density matrix corresponding to ψ has the elements

$$\rho_{ij} = c_i \bar{c}_j. \tag{2.31}$$

Hence, for the pure state ψ, the basic probabilistic postulate of quantum mechanics, (2.28), has the form

$$P(a_j) = \rho_{jj} = c_j \bar{c}_j = |c_j|^2. \tag{2.32}$$

This postulate can be written without using the coordinates of the state vector ψ with respect to the basis of eigenvectors of a quantum observable. We remark that, since the basis of eigenvectors of a Hermitian operator can always be selected as orthonormal, the coordinates c_j can be expressed in the form: $c_j = \langle \psi, e_j \rangle$. Hence, the Born's rule takes the form:

$$P(a_j) = |\langle \psi, e_j \rangle|^2. \tag{2.33}$$

Projection Postulate; Resolution of Uncertainty

The next natural question is about the post-measurement state. What will happen with the state ψ after a measurement? By the von Neumann *projection postulate* the superposition (2.29) is reduced to just one term, the state e_j corresponding to the eigenvalue a_j obtained in the measurement. We remark that we consider the simplest case: all eigenvalues are different from each other; the case of nondegenerate spectrum.

This procedure can be interpreted in the following way:

This superposition encodes the uncertainty in the results of measurements for the observable A. *Roughly speaking, before a measurement a quantum system* "*d*oes not know how it will answer to the question A." *The mathematical expressions* (2.29) *and* (2.32) *encode potentialities for different answers. Thus a quantum system in the superposition state* ψ *does not have any value of* A *as its objective property.*[3] *After a measurement the superposition is reduced to just one term in the expansion* (2.29) *corresponding the value of* A *obtained in the process of measurement.*

We remark that the state reduction is often called state *collapse.* Some experts in quantum foundations treat superposition physically and not simply operationally; for them, the collapse is also a physical event.[4] We state again that in this book we proceed with the operational interpretation of the quantum formalism. By this interpretation superposition (2.29) expresses the uncertainty in the expected results of the A-measurement. And nothing more! When the result a_j is detected, this uncertainty is resolved. Hence, "collapse" takes place not in the physical space, but in the information space.

Encoding of the uncertainty in bio-systems by superpositions is one of the cornerstones of quantum bio-informatics. A bio-system as well as a quantum physical system can be in a state of uncertainty on possible reactions to measurements. Such states are *mathematically encoded* as linear superpositions. Measurement resolves such superpositions.

---

[3] See Sect. 1.3 for a further discussion.

[4] In particular, by the orthodox Copenhagen interpretation ψ is interpreted as the physical state of a system. As a consequence, its collapse is a physical event.

Degenerate Eigenvalues

Consider now the general case: the eigenvalues can be degenerate and eigensubspaces[5] need not be one dimensional. Suppose that by measuring an observable represented by the Hermitian operator A its eigenvalue a was obtained; denote the projector to this eigensubspace by π_a. Then by the projection postulate the input pure state ψ is transformed into (again pure) state

$$\psi_{\text{out};a} = \frac{\pi_a \psi}{\|\pi_a \psi\|}. \tag{2.34}$$

We remark that although this is the standard definition used in modern quantum theory, von Neumann distinguished sharply the cases of non-degenerate spectrum (i.e., in the finite dimensional case, all eigenvalues are different) and degenerate spectrum. In the latter case the definition (2.34) is due to Lüders. Originally von Neumann assumed that a measurement can transfer a pure state into a mixed state given by a density operator, see [3]. However, we would not disturb biologists by such foundational problems. Hence, we proceed with the Lüders form of the projection postulate even in the case of degenerate spectrum (as the simplest form of the projection postulate).[6] If the input state is given by the density operator ρ, then

$$\rho_{\text{out};a} = \frac{\pi_a \rho \pi_a}{\text{tr}\pi_a \rho \pi_a}. \tag{2.35}$$

Suppose now that by measuring A we are interested in the state (mixed) representing the ensemble of states of all systems after measurement, so this ensemble is the mixture of the ensembles corresponding to the results $A = a_j$ for different a_j. Then this state is mathematically represented by the density operator

$$\rho_{\text{out}} = \sum_a \pi_a \rho \pi_a. \tag{2.36}$$

It can be represented as a mixture of the states $\rho_{\text{out};a}$:

$$\rho_{\text{out}} = \sum_a P_a \rho_{\text{out};a}, \text{ where } P_a = \text{tr}\pi_a \rho \pi_a. \tag{2.37}$$

Self-measurements

We stress that in quantum bio-informatics we interpret the notion of measurement in a more general way than in quantum physics. In particular, measurements can be *self-measurements*, see Sect. 1.3; we also treat *decoherence* as a form of measurement.

---

[5] The set of all the eigenvectors is a linear subspace of $\mathscr{H}$.

[6] Even the majority of physicists have never read the von Neumann's book [3] and they have no idea that von Neumann distinguished degenerate and non-degenerate cases. We are aware that this distinction may play an important role in biology.

Decoherence is the process of transformation of superposition (2.29) into the diagonal density matrix, $\rho = \text{diag}(\rho_{11}, \ldots, \rho_{mm})$ corresponding to the total resolution of uncertainty.[7]

Self-measurements play an important role in the brain science, cognitive science, psychology. Decoherence will be applied to describe processes of decision making (in cognitive psychology and game theory, Chap. 4) and in modeling of biological evolution, including evolution on the cellular level, Chap. 8.

2.2.3 Dirac Notation

Dirac was one of the inventors of the quantum formalism [5]. He introduced his own symbolic notation for the Hilbert space linear algebra which was not used in mathematics. This notation became very common in quantum physics and especially in quantum information theory. We shall now present shortly Dirac's symbolic notation. In this books we shall use it very often (but not always).

Consider an observable A, denote its eigenvalues by $a_1, \ldots, a_m$ and the corresponding eigenvectors by $e_1, \ldots, e_m$. Suppose again that all eigenvalues are different. Then by Dirac e_j is denoted as $|a_j\rangle$. Even an arbitrary pure state ψ is often written as $|\psi\rangle$ (just for convenience). In Dirac's notation superposition (2.29) is written as

$$|\psi\rangle = c_1|a_1\rangle + \cdots + c_m|a_m\rangle, \tag{2.38}$$

or simply $|\psi\rangle = \sum_a c_a|a\rangle$. (We repeat that this is just symbolic expression for the operator with the set of eigenvectors $|a\rangle$.) The sum "can be continuous", i.e., it will be an integral representation (in the case of observables with continuous spectra such as e.g. position and momentum of a quantum particle): $|\psi\rangle = \int c(a)|a\rangle da$, where $\int |c(a)|^2 da = 1$.

The Born's rule (2.33) is now written as

$$P(a_j) = |\langle\psi|a_j\rangle|^2. \tag{2.39}$$

Consider some pure state $|\psi\rangle$, then the symbol $|\psi\rangle\langle\psi|$ denotes the operator of orthogonal projection onto the vector $|\psi\rangle$. Thus the operator A with the system of distinct eigenvalues $a_1, \ldots, a_m$ can be written as

$$A = \sum_{i=1}^{m} a_i|a_i\rangle\langle a_i|,$$

[7] Decoherence is a complicated interpretational issue of quantum mechanics. Some (but not all) researchers treat decoherence as a form of measurement.

or simply $A = \sum_a a|a\rangle\langle a|$. We shall come back to the problem of encoding of quantum information in Sect. 2.5.3, after the introduction of state spaces of compound systems (given by tensor products of Hilbert state spaces of subsystems).

2.2.4 Qubit

The basic notion of quantum information theory is quantum bit (qubit), the quantum analog of classical bit of information. Consider any dichotomous observable $A = a_1, a_2$ with $a_1 < a_2$, in particular, it acts in the two dimensional Hilbert space. Let us encode 0, 1 by its eigenvectors. We can always calibrate the pointer of the corresponding measurement device in such a way that $a_1 = 0$ and $a_2 = 1$, the encoding rule has the form $0 \to |0\rangle$, $1 \to |1\rangle$. The crucial point is that, besides the states $|\alpha\rangle, \alpha = 0, 1$, a quantum system can be in superposition of these states,

$$|\psi\rangle = c_0|0\rangle + c_1|1\rangle, \qquad |c_0|^2 + |c_1|^2 = 1. \tag{2.40}$$

Thus a single quantum system can carry with its state not just either 0 or 1, i.e., one bit of information, but both 0 and 1, with some probabilistic weights. However, by measurement of A it is possible to extract just one bit of information by getting either the result $A = 0$ or $A = 1$. The crucial point is that (2.40) represents not simply the classical probability distribution with two weights $p_0 = |c_0|^2$, $p_1 = |c_1|^2$. Qubit can generate interference effects, but the classical probability distribution not. We shall discuss this problem in more detail later.

2.2.5 Wave Function

Up to now we have restricted the presentation to the finite dimensional case. In real physics the Hilbert space of quantum states is *infinite dimensional*, the basic example is given by the space of square integrable complex valued functions, $x \to \psi(x)$, where x is the spatial variable, i.e., functions such that $\int |\psi(x)|^2 dx < \infty$. We denote this space by the symbol $L^2(\mathbb{R}^n)$, where $x \in \mathbb{R}^n$. A pure state ψ, i.e., a function such that

$$\int |\psi(x)|^2 dx = 1, \tag{2.41}$$

is known as a *wave function*.

For the position observable, the spatial variable x plays the role of the discrete index j in (2.32) (as well as in (2.29), (2.30)). Since the dimension of state space is infinite, eigenfunctions are in general generalized, i.e., they do not belong to the state space. A larger space has to be in use. In our case such a larger space containing eigenfunctions is some space of distributions (generalized functions). For the position operator, the (generalized) eigenfunction corresponding to the (generalized) eigenvalue x_0 is given by the Dirac δ-function whose argument is shifted by x_0, i.e.,

$e_{x_0}(x) = \delta(x - x_0)$. Thus eigenfunctions $e_{x_0}(x)$ are "enumerated" with the aid of the continuous index x_0. And formally any square integrable function can be represented as the superposition of eigenfunctions of the position operator:

$$\psi(x) = \int \psi(x_0) e_{x_0}(x) dx_0. \tag{2.42}$$

This is a continuous analog of superposition (2.29). A continuous analog of the normalization condition (2.30) is given by the equality (2.41). For a pure state, the probabilistic postulate of quantum mechanics has the form:

$$P(x) = |\psi(x)|^2, \tag{2.43}$$

"the probability to find a quantum system with the wave function ψ at the point x is equal to the squared absolute value of this function at this point." This is the original formulation of the basic probabilistic postulate of quantum mechanics, *Born's rule*. This is a postulate; it cannot be derived in the conventional quantum theory; however, cf. [6].

In the real quantum physics state spaces are always of the L^2-type, i.e., they are infinite-dimensional. However, in quantum information one typically selects just a few degrees of freedom, which are then used for the representation of information and proceed in finite dimensional state space. In this book we shall apply the same strategy. However, physical state space and wave function will be used in Sect. 4.1.1, which is devoted to the presentation of interference of photons.

2.3 Schrödinger Dynamics and Its Role in Quantum Bio-informatics

We can assume that, like in a quantum physical system, the evolution of the QL-state of a bio-system isolated from the environment is described by the *Schrödinger equation* (with a minor modification related to the usage of the Planck constant in physics):

$$i\gamma \frac{d\psi}{dt} = H\psi(t), \qquad \psi(0) = \psi^0. \tag{2.44}$$

Instead of the Planck constant h, we use a constant γ, which has the dimension of time. We recall that the *Planck constant* has the dimension of action, i.e., energy $\times$ time. Here H is a Hermitian operator acting in the space of QL-states of bio-systems. In quantum physics $\mathscr{H}$ is the energy operator; it is called *Hamiltonian*. The energy dimension of this operator matches with the action dimension of the Planck constant. In quantum bio-informatics this operator does not represent the physical energy of a bio-system. (We state again that quantum bio-informatics is not at all quantum biology.) It is the generator of purely information dynamics. It is convenient to have it dimensionless and this motivates the usage of the time scale constant γ, instead of the action constant h. In coming QL-models γ gives the time scale of information processing in biological systems. We do not claim that there exists some universal constant scaling dynamics in all bio-systems. Each population of bio-systems can have its is own temporal scale.

We shall also call the generator of the Schrödinger dynamics Hamiltonian (the operator H) and we hope that it will not cause misunderstanding.

The Schrödinger equation is, in fact, a system of linear differential equations with complex coefficients; in the one dimensional case $\mathscr{H}$ is just a real number and the general solution has the form of the imaginary exponent: $\psi(t) = e^{\frac{-it\mathscr{H}}{\gamma}} \psi^0$. In the general case $\mathscr{H}$ is an operator and the solution is represented in the form of imaginary operator-exponent (for the fixed basis it is simply the exponent of the matrix):

$$\psi(t) = U_t \psi^0, \; U_t = e^{\frac{-it\mathscr{H}}{\gamma}}. \tag{2.45}$$

As well as the one dimensional imaginary exponent, the operator-exponent describes *oscillating dynamics.* It is more complicated than in the one dimensional case; it is a mixture of many oscillating imaginary exponents. For us, the most important point is that such a dynamics will never stabilize to any definite solution.

We point to the following fundamental property of the Schrödinger dynamics. The evolution operator U_t, see (2.45), is a *unitary operator*, i.e., it preserves the scalar product:

$$\langle U_t \psi | U_t \psi \rangle = \langle \psi | \psi \rangle. \tag{2.46}$$

Thus this dynamics transfers a pure quantum state to another pure quantum state.

In quantum physics, the Schrödinger equation plays a crucial role. It seems not to be the case in biology. Here it is very difficult and practically impossible to provide the complete isolation from environment. Take a single cell and isolate it from other cells. If we still provide the supply of chemical staffs, this cell will survive, but its behavior in the absence of signaling from other cells will be very artificial. For example, such an important process as cell differentiation will be impossible. If we try to isolate a cell from the supply of all chemical staffs, it will die. The same can be said about the isolated (mentally or/and physically) brain.

Since any general quantum state, a density operator, can be represented as a mixture of density operators corresponding to pure states, the Schrödinger dynamics for pure states implies the following dynamics for density operators:

$$\gamma \frac{d\rho}{dt}(t) = -i[H, \rho(t)], \quad \rho(0) = \rho^0. \tag{2.47}$$

(In quantum physics the Planck constant h is used instead of the time scaling factor γ). This equation is known as the *Von Neumann equation* [3]. By using representation (2.45) of the Schrödinger evolution for the pure state we represent the evolution of the density operator ("mixed state") in the form

$$\rho(t) = U_t^* \rho^0 U_t, \tag{2.48}$$

where, for an operator W, the symbol W^* denotes its *adjoint operator*. The latter is defined by the equality

$$\langle W\psi_1|\psi_2\rangle = \langle\psi_1|W^*\psi_2\rangle; \tag{2.49}$$

by denoting the matrix elements of these operators as w_{ij} and w_{ij}^* we have $w_{ij}^* = \bar{w}_{ji}$.

2.4 Theory of Open Quantum Systems in Biology

What kind of the mathematical apparatus can be used to model adaptive biological phenomena operationally? An important class of adaptive dynamical systems can be described by the apparatus of the theory of *open quantum systems,* see, e.g., Ingarden et al. [7], Ohya and Volovich [8]. In complete accordance with the operational viewpoint to the formalism of quantum physics, we can apply the theory of open quantum systems in biology, from genetics and cellular biology to the brain science, cognitive science and psychology. By the theory of open quantum systems, the dynamics of the QL-state of a bio-system interacting with an environment is approximately described by *quantum master equation*, the Gorini-Kossakowski-Sudarshan-Lindblad (GKSL) equation, see, e.g., [7].[8]

In cellular QL-models at the beginning of interaction with an environment the QL-state of a population of cells is characterized by a *high degree of uncertainty about possible changes* (cf. Sect. 2.2.2), which can be generated via the coupling with an environment. The quantum master equation describes the process of resolution of this state of uncertainty and approaching the complete matching with the environment, Sect. 2.4. This process can be considered as *decoherence of the QL-state* through interaction with an environment. As a result, the state loses its fundamentally quantum(-like) feature, *superposition* of a few alternatives, and the final situation can be described by classical probability theory. Mathematically, this is the process of approaching a *steady state* solution of the quantum master equation.

The first application of the theory of open quantum systems to biology was in the domain of cognitive psychology and game theory, see [9, 10]. The classical probability distribution corresponding to the steady state was used to model decision making by cognitive systems; in particular, by gamblers in games of the Prisoner's Dilemma type; cf. with QL-modeling performed on the basis of Schrödinger equation [11, 12].

One of the novel proposals realized in this book is *to use the machinery of the theory of open quantum systems and to describe dynamics of the QL-state of a population of bio-systems* by using the quantum master equation (the GKSL-equation). We state again that this equation can be used to describe transitions from states

[8] In quantum information theory this equation is often referred as Lindblad equation.

of uncertainty given by superpositions to classical probability distributions. Hence, such an equation cannot be an equation with respect to the pure state represented as a *vector* belonging to complex Hilbert space and normalized by one (We remark that the Schrödinger equation is an equation with respect to the pure states!). Pure states represent superpositions of possibilities which have to disappear at the end: when a bio-system will make its decision.[9] We have to use general quantum states represented by *density operators*. As was stressed in Sect. 2.2, initial pure quantum states (superpositions describing uncertainty) and the final classical probability distribution can be represented by density operators. (Although the picture of a population of bio-systems making decisions is very illustrative in the brain science, cognitive science and psychology, in general we use the operational interpretation: the output (classical) probabilities give us the probability distribution of decisions.)

In the quantum Markovian approximation, the dynamics of the state of a system interacting with an environment is described by the GKSL-equation (again with a minor modification related to the usage of the Planck constant in quantum physics):

$$\gamma \frac{d\rho}{dt} = -i[H, \rho(t)] + \mathscr{W}\rho(t), \rho(0) = \rho^0, \tag{2.50}$$

where H is a Hermitian operator determining the internal dynamics of the QL-state of a population of bio-systems isolated from the environmental pressure ("cell's Hamiltonian") and the linear operator $\mathscr{W}$ describes the environmental pressure.

Opposite to H (which is simply a Hermitian operator; for a fixed basis it is represented by a Hermitian matrix), in general the operator $\mathscr{W}$ has a complex mathematical structure. It has such a form that starting with a density operator ρ° we shall get density operators at all instances of time. For a moment, the specific structure of $\mathscr{W}$ is not important for us; see, e.g., [7, 8], and Sect. 8.2.3 for mathematical details. Biologically this operator is determined by the properties of the environment, including the initial state of the environment.

Here γ is the time scale constant, it determines the temporal scale of the biological dynamics. By using such a scaling factor of the dimension of time, we are able to proceed with dimensionless Hamiltonian H and the environmental operator $\mathscr{W}$.

We state once again that for our QL-modeling it is crucial that, for a very general class of GKSL-equations, the environmental operator $\mathscr{W}$ drives (in the limit $t \to \infty$) the QL-state of a biological population, $\rho(t)$, to the *steady solution*: $\rho(t) \to \rho_{st}$. Typically the uncertainty (in the form of superposition) is eliminated from the asymptotic state ρ_{st}.

In our QL-model such a steady state is considered as the result of the biological dynamics in the environment (mathematically represented by the operator $\mathscr{W}$). For example, we can consider *epigenetic evolution* (Chap. 8). Here the limiting probability distribution $\rho_{epi;st}$ describes the probability distribution of epimutations which took place in a cell population as a consequence of interaction with the environment.

[9] We treat the notion of decision very generally: from decisions made by people to "cell's decisions", e.g., to undergo epimutation.

Internal uncertainty, to epimutate or not epimutate, was resolved and a stable *phenotype* was created.

Finally, we remark that under natural restrictions a selection operator produces the *same steady state for all possible initial states,* Sect. 8.3.1. Such an open system dynamics simulates, for example, the spreading of mutations or epimutations in a population. The pressure of the environment can be so strong that the same genotype or phenotype is created independently of the initial states of populations. Such a dynamics can also have applications in social science and political technologies (See, e.g., [13, 14] for modeling of dynamics of *party governance* in the USA political system. Here a QL-model of decision-making of American voters was elaborated. In the framework of open quantum systems a possibility of driving populations of voters to the fixed state, e.g., to vote for republicans independently of their initial preferences was modeled.)

We shall continue our discussion on the theory of open quantum systems and the theory of adaptive quantum dynamics in Sect. 2.6 after the introduction to the theory of compound quantum systems.

2.5 Compound Systems

The quantum description of a compound system S consisting of two subsystems S_1 and S_2 with state (Hilbert) spaces $\mathcal{H}_1$ and $\mathcal{H}_2$, respectively, is based on the representation of states of such a system in the tensor product space $\mathcal{H} = \mathcal{H}_1 \otimes \mathcal{H}_2$. Since the notion *tensor product* is not used so much in biology,[10] in the following sections we present briefly the construction of Hilbert space $\mathcal{H}_1 \otimes \mathcal{H}_2$. Although in quantum information theory and in this book we shall use the formal algebraic definition, which is especially useful for finite dimensional Hilbert spaces, we prefer to start with the construction originally used by Von Neumann [3], namely, the tensor product of two spaces of square integrable functions.

Tensor product of functional spaces. Let now both state spaces be L^2-spaces, $\mathcal{H}_1 = L^2(\mathbb{R}^n)$ and $\mathcal{H}_2 = L^2(\mathbb{R}^m)$, see Sect. 2.2.5. Take two functions; $\psi \equiv \psi(x)$ belongs to $\mathcal{H}_1$ and $\phi \equiv \phi(y)$ belongs to $\mathcal{H}_2$. By multiplying these functions we obtain the function of two variables $\Psi(x, y) = \psi(x) \times \phi(y)$. It is easy to check that this function belongs to the space $\mathcal{H} = L^2(\mathbb{R}^{n+m})$. Take now n functions, $\psi_1(x), \ldots, \psi_n(x)$, from $\mathcal{H}_1$ and n functions, $\phi_1(y), \ldots, \phi_n(y)$, from $\mathcal{H}_2$ and consider the sum of their pairwise products:

$$\Psi(x, y) = \sum_i \psi_i(x) \times \phi_i(y). \tag{2.51}$$

[10] In the coordinate form tensor products of vectors and matrices are also known under the name *Kronecker product*. This structure is widely used in various computational algorithms including computational biology.

This function also belongs to $\mathscr{H}$.

It is possible to show that any function belonging to $\mathscr{H}$ can be represented as (2.51), where the sum is in general infinite. Multiplication of functions is the basic example of the operation of the tensor product. The latter is denoted by the symbol $\otimes$. Thus in the example under consideration $\psi \otimes \phi(x, y) = \psi(x) \times \phi(y)$. The tensor product structure on $\mathscr{H}_2 = L^2(\mathbb{R}^m)$ is symbolically denoted as $\mathscr{H} = \mathscr{H}_1 \otimes \mathscr{H}_2$.

Consider now orthonormal bases in $\mathscr{H}_k$, $(e_j^{(k)})$, $k = 1, 2$. Then (functions) $(e_{ij} = e_i^{(1)} \otimes e_j^{(2)})$ form an orthonormal basis in $\mathscr{H}$: any $\Psi \in \mathscr{H}$, can be represented as

$$\Psi = \sum c_{ij} e_{ij} \equiv \sum c_{ij} e_i^{(1)} \otimes e_j^{(2)}, \tag{2.52}$$

where

$$\sum |c_{ij}|^2 < \infty. \tag{2.53}$$

Those who work with electromagnetic signals in biology, e.g., in the brain research, have experience in expanding electromagnetic signals with respect to various bases, e.g., using the Fourier expansion or the wavelet expansion. Some bases are indexed by continuous parameters; integrals take place of sums. Thus the notion of basis in the L^2-space is widely known. However, there is a crucial difference between the classical field and quantum mechanical representations of compound systems. The state of a classical bi-signal consisting of two components is represented in the *Cartesian product* of the corresponding L^2-spaces. And the state of a quantum bi-system, e.g., bi-photon, is represented in the tensor product space (Sect. 2.5.1). One may state that the crucial difference between the classical and quantum physical models is in the representation of states of compound systems. (Although, as we have already seen in previous sections, the descriptions of non-compound systems differ essentially.)

Tensor product, the algebraic definition. Consider now two finite dimensional Hilbert spaces, $\mathscr{H}_1, \mathscr{H}_2$. For each pair of vectors $\psi \in \mathscr{H}_1, \phi \in \mathscr{H}_2$, we form a new formal entity denoted by $\psi \otimes \phi$. Then we consider the sums $\Psi = \sum_i \psi_i \otimes \phi_i$. On the set of such formal sums we can introduce the linear space structure. (To be mathematically rigorous, we have to constraint this set by some algebraic relations to make the operations of addition and multiplication by complex numbers well defined). This construction gives us the tensor product $\mathscr{H} = \mathscr{H}_1 \otimes \mathscr{H}_2$. In particular, if we take orthonormal bases in $\mathscr{H}_k$, $(e_j^{(k)})$, $k = 1, 2$, then $(e_{ij} = e_i^{(1)} \otimes e_j^{(2)})$ form an orthonormal basis in $\mathscr{H}$, any $\Psi \in \mathscr{H}$, can be represented as (2.52) with (2.53).

The latter representation gives the simplest possibility to define the tensor product of two arbitrary (i.e., may be infinite-dimensional) Hilbert spaces as the space of formal series (2.52) satisfying the condition (2.53).

Besides the notion of tensor product of states, we shall also use the notion of tensor product of operators. Consider two linear operators $A_i : \mathscr{H}_i \to \mathscr{H}_i, i = 1, 2$. Their tensor product $A \equiv A_1 \otimes A_2 : \mathscr{H} \to \mathscr{H}$ is defined starting with the tensor products of two vectors: $A\psi \otimes \phi = (A_1\psi) \otimes (A_2\phi)$. Then it is extended by linearity.

By using the coordinate representation (2.52) the tensor product of operators can be represented as

$$A\Psi = \sum c_{ij} A e_{ij} \equiv \sum c_{ij} A_1 e_i^{(1)} \otimes A_2 e_j^{(2)}, \tag{2.54}$$

If the operators $A_i, i = 1, 2$, are represented by matrices (with respect to the fixed bases), i.e., $A_i = (A_{kl}^{(i)})$, then the matrix $A = (A_{kl.nm})$ with respect to the tensor product of these bases can be easily calculated.

In the same way one defines the tensor product of Hilbert spaces, $\mathcal{H}_1, \ldots, \mathcal{H}_n$, denoted by the symbol $\mathcal{H} = \mathcal{H}_1 \otimes \cdots \otimes \mathcal{H}_n$. We start with forming the formal entities $\psi_1 \otimes \cdots \otimes \psi_n$, where $\psi_j \in \mathcal{H}_j, j = 1, \ldots, n$. Tensor product space is defined as the set of all sums $\sum_j \psi_{1j} \otimes \cdots \otimes \psi_{nj}$ (which has to be constrained by some algebraic relations, but we omit such details). Take orthonormal bases in $\mathcal{H}_k$, $(e_j^{(k)})$, $k = 1, \ldots, n$. Then any $\Psi \in \mathcal{H}$ can be represented as

$$\Psi = \sum_\alpha c_\alpha e_\alpha \equiv \sum_{\alpha=(j_1 \cdots j_n)} c_{j_1 \cdots j_n} e_{j_1}^{(1)} \otimes \cdots \otimes e_{j_n}^{(n)}, \tag{2.55}$$

where $\sum_\alpha |c_\alpha|^2 < \infty$.

2.5.1 Entanglement

Consider a compound quantum system $S = (S_1, S_2)$, where the subsystems $S_i, i = 1, 2$, have state spaces $\mathcal{H}_i$. Then the state space of S is $\mathcal{H} = \mathcal{H}_1 \otimes \mathcal{H}_2$. States and observables are defined as it was presented in Sect. 2.2. However, the presence of the tensor product structure generates a new important notion, namely, *entanglement of states*. Consider the case of two uncorrelated systems S_1 and S_2 in states ψ_1 and ψ_2. Then the state of the compound system S is given by the tensor product

$$\Psi = \psi_1 \otimes \psi_2. \tag{2.56}$$

(In particular, for wave functions we have $\Psi(x, y) = \psi_1(x) \times \psi_2(y)$.) However, if there are some correlations between the degrees of freedom in S_1 and S_2, then the state of S cannot be represented in the form (2.56). Such a state is called an *entangled state*. The notion of entanglement can be generalized to nonpure states (which are represented by density operators acting in $\mathcal{H}$.) The interpretation of this notion is still the topic of hot debates [4]. There can be found variety of interpretations: from the original interpretation of Einstein et al. [15], who considered entanglement as the Hilbert space representation of classical correlations between systems to modern interpretations such as the subjective probability interpretation – entanglement as the correlation of probabilistic knowledge about subsystems in the mind of an observer, e.g., Fuchs and coauthors [16–18]. In this book we shall not try to select one of possible interpretations of entanglement. For the operational approach, we are fine even with Fuchsian interpretation. However, we cannot reject a possibility of classical probabilistic interpretation of entanglement, see [6, 19–24]. In quantum bio-informatics we shall simply use the tensor product representation (and, hence, entanglement) of states for compound bio-systems. For example, this representation plays an important role in modeling correlations between the expressions of different genes in genome as well as correlations between mutations (as well as epimutations) in different genes, Chap. 8. In the theory of decision making we shall consider entanglement between states of different parties (typically named Alice and Bob), Chap. 4.

Is the usage of tensor product representation and entanglement biologically justified? It seems that this is not a proper question. In quantum physics we cannot find "real physical reasons" for the usage of the aforementioned mathematical constructions. They are fruitful, because they work well for the representation of statistical experimental data. In quantum bio-informatics we shall proceed in the same way. We remark that in the abstract algebraic framework elements of tensor products are defined in a formal way; there is no heuristic coupling between Ψ belonging to $\mathscr{H} = \mathscr{H}_1 \otimes \mathscr{H}_2$ and its components, e.g., (2.52), belonging to $\mathscr{H}_i$. In the operational approach this situation is natural. We just need a symbolic representation of states of compound quantum systems.[11]

2.5.2 *Tensor Products and Contextuality of Observables for a Single System*

Tensor product structure can appear not only in the description of a state space of a compound system, but even in the case of a single system. Sometimes it is possible to factorize the state space $\mathscr{H}$ of a single quantum system, say a neutron, into the tensor product $\mathscr{H} = \mathscr{H}_1 \otimes \mathscr{H}_2$ in such a way that two compatible observables (Hermitian operators A_1, A_2 acting in $\mathscr{H}$) can be represented in the form $A_1 = a_1 \otimes I$ and $A_2 = I \otimes a_2$, where $a_i, i = 1, 2$, are Hermitian operators in $\mathscr{H}_i$ and I denotes the unit operator, $Ix = x$ (For neutrons, A_1 and A_2 can be selected as the position and spin observables.) Here we also can consider entangled states. However, this is *entanglement of observables* for a single system and not states of a few systems. This approach is very useful for quantum bio-informatics. For example, we can consider the entanglement of various genetic or epigenetic markers of a single cell

[11] "What is beyond this symbolism?"—this is a separate question (Sect. 9.2; see also [19]). In a series of papers [6, 20, 21, 23, 24] quantum systems were represented by classical random fields. In this approach elements of the tensor product can be visualized via the functional representation.

(Sect. 8.5). Using of the entanglement (-like)[12] model can explain the high speed of the epigenetic evolution, in which inheritable epigenetic markers can be created during only one generation of a cellular population.

2.5.3 A Few Words About Quantum Information

The notion of qubit was introduced in Sect. 2.2.4. Now we consider n-qubit states. One qubit space is two dimensional with the basis encoding 0 and 1: $|0\rangle, |1\rangle$. It is isomorphic to $\mathbb{C}^2$. Now we consider the tensor product of n qubit spaces. The basis in this space has the form $|x\rangle \equiv |x_1\rangle \otimes \cdots \otimes |x_n\rangle$, where $x_j = 0, 1$. Thus each basis vector $|x\rangle$ encodes the string of zeros and ones of the length n. As in the one dimensional case, it is possible to form weighted superpositions of these basis vectors. These are quantum n-bit states. The dimension of the n-qubit state space is equal to 2^n. This space is isomorphic to $\mathbb{C}^{2^n}$. Typically one omits the signs of the tensor product and writes $|x\rangle \equiv |x_1 \cdots x_n\rangle$. Thus a general n-qubit state has the form

$$|\psi\rangle = \sum_x c_x |x\rangle, \text{ where } \sum_x |c_x|^2 = 1. \tag{2.57}$$

The main distinguishing feature of quantum information theory is that the dimension of a state space increases exponentially with the linear increasing of the number n of subsystems of a compound system or more generally entangled degrees of freedom (may be even of a single system, see Sect. 2.5.2). This is the main computational resource. And it is huge. A relatively small physical system (say $n = 100$) can process a gigantic information state (in our example, the superposition of 2^n bits of information). However, as in the single qubit case, this huge information resource is unapproachable: a measurement can give us only one result.

Quantum computations are based on the physical realization of the Schrödinger dynamics (unitary transformations) in the n-qubit spaces and the final measurement. The last step has to be designed in an intelligent way, since the solution of a problem has to be found in a single basis state obtained via measurement (Lüders projection). This is one of the reasons why only a special class of problems can be solved by using quantum computations. This class is very restricted. Ohya et al. proposed to combine quantum algorithms with classical chaotic dynamics [25, 26]. This approach

[12] We remark that information features of entanglement can be modeled (mimic) by using coarse graining procedure for classical stochastic processes, even for classical Brownian motion [19]. In the latter paper the entanglement is exhibited at the level of observables corresponding to coarse graining.

provides a possibility to solve new NP-problems[13], e.g., the SAT-problem, see Ohya and Volovich [8, 27, 28].

However, in general the usage of unitary transformations makes physical realization of quantum algorithms very difficult, since unitary dynamics is possible only in isolation from the environment. The latter is practically impossible. The entangled quantum n-qubit states are subjected to quick decoherence (destruction of superposition (2.57) and spontaneous reduction to one of the basis states). There is simply not enough time to perform quantum calculations. Therefore we do not think that bio-systems can perform real quantum calculations based on unitary transformations. Our conjecture is that bio-systems can process entangled states, but by using non-unitary dynamics of the GKSL-type (Sect. 2.4), see Sect. 8.4.2 for more detail.

2.6 From Open Quantum System Dynamics to State-Observable Adaptive Dynamics

In Sect. 2.4 we considered the dynamics of the state of a system, say S, interacting with an environment. To derive this dynamics, the Eq. (2.50), one proceeds in the following framework. The states of the system under consideration are represented in a Hilbert space $\mathcal{H}$ and the states of an environment (bath) are represented in another Hilbert space $\mathcal{K}$. Since an environment is a huge system, its state space has very high dimension, in a mathematical model it can be considered as infinite-dimensional. The states of the compound system, the system S and the environment, are represented in the Hilbert space $\mathcal{H} \otimes \mathcal{K}$. In the theory of open quantum systems the evolution of a pure state of such a compound system is described (in the complete accordance with the postulates of quantum mechanics) by the Schrödinger equation. (In quantum bioinformatics we use its slight modification by using the dimensionless Hamiltonian and the time scaling constant γ, instead of the Planck constant, see Sect. 2.3. In this section we also proceed with such γ. In physics one can, as usual, proceed with Hamiltonians having the dimension of physical energy and the Planck constant instead of the scaling factor γ.) The Hamiltonian of a compound system is in general very complicated. Therefore in real calculations its various approximations are in use. Set $U_t = e^{-itH/\gamma}$, the evolutionary operator of S interacting with the environment. Then unitary dynamics of the pure state is given as

$$\Psi(t) = U_t \Psi_0, \tag{2.58}$$

[13] In computability theory, a decision problem, which has two possible answers, "yes" or "no", for an input of question is studied well, and its complexity is classified into several complexity classes. NP (Nondeterministic Polynomial time) problem is a decision problem, whose solution is not given in polynomial time on a non-deternimistic Turing machine, and NP-complete problem is a NP problem reduced in polynomial time from any other NP problems. Whether NP-complete problem can be reduced to Polynomial problem is one of the millennium prize problems, it has been discussed for thirty yeas.

where Ψ_0 is the initial state of S combined with the environment. For general states given by density operators, the dynamics is given by the von Neumann equation and, hence, it can be represented in the form, see (2.48),

$$R(t) = U_t R_0 U_t^*, \tag{2.59}$$

where R_0 is the initial state of the compound system. Typically one is interested only in the dynamics of S (and not interested in what happens with an environment around S). This dynamics is obtained by taking the trace of the density operator of the compound system with respect to all environmental degrees of freedom:

$$\rho(t) = \mathrm{tr}_{\mathscr{K}} R(t) = \mathrm{tr}_{\mathscr{K}} U_t R_0 U_t^* \tag{2.60}$$

where $\mathrm{tr}_{\mathscr{K}}$ denotes the partial trace with respect to the subspace $\mathscr{K}$ of the complete state space $\mathscr{H} \otimes \mathscr{K}$.

This dynamics is very complex because of the complexity of Hamiltonian of the total system and, in particular, the complexity of interaction between S and environment. Then under a number of assumptions it is possible to prove that the dynamics given by (2.60) can be reduced to the quantum master Eq. (2.50). One of the main assumptions is Markovness of dynamics. Another important assumption is that interaction of S with the environment cannot change the state of the environment (or if it changes then this change can be considered as negligible). (In Sect. 8.2.2 we shall discuss conditions of derivation of (2.50) from (2.60) in more detail and in the biological framework; in application to biological evolution.) We emphasize that in general dynamics (2.60) is non-Markovian. In the latter case the system S "remembers" the states of the compound system, S and the environment for sufficiently long period and it makes its state update on the basis of this information.

We think that to match completely with biology, the theory of open quantum systems has to be generalized. First of all we are critical to the approach assuming a possibility to construct interaction Hamiltonian. This is very difficult in physics. And in biology this problem is even more complicated. It is difficult or may be even impossible to describe formally the character of the interaction of a bio-system S with its environment. Such an environment has not only physical and chemical counterparts, but also a complex information component. And the latter plays the crucial role in many biological processes (e.g., signaling of surrounding cells is crucial in the process of cell differentiation, Chap. 8. Starting with the Hamiltonian, one postulates the Schrödinger dynamics for pure states and then the von Neumann dynamics is its trivial consequence. However, the assumption that such a complex system as an environment interacting with a system S can be in a pure state has no justification, especially in biology. It is more natural to assume the presence of mixture of states. Of course, one can easily avoid the last problem by postulating directly the von Neumann equation. However, even for a density operator, the usage of the *unitary dynamics* (2.59) is not justified.)

We think that, especially in quantum bio-informatics, the direct usage of the quantum Hamiltonian approach to describe the dynamics of a system in some environment is not justified. Moreover, a bio-system is able to perform measurements, both on surrounding environment and on itself (self-measurements, Sect. 2.2.2). Such measurements are not described by unitary operators. Since environment information component is typically based on the activity of other biological systems, e.g., cells surrounding the cell under consideration, we cannot assume that by testing the environment a bio-system does not change the state of this environment. Therefore, instead of tracing of the unitary dynamics of the compound system, a system S interacting with an environment, see (2.60), and then approximating it in some way (Markovian or even non-Markovian), we develop a new theory—the theory of *state-observable adaptive dynamical systems*, Chap. 4.

By this theory *a system S (e.g., a cell) permanently updates its state ρ by performing measurements on the state of the surrounding environment (hence, the system state update depends on the state σ of the environment and measurement performed by S) and successive self-measurement.* This state update cannot be done without changing the state of the environment. Therefore formally we can describe both measurement on the environment and self-measurement of S as a measurement, say Q, on the compound system, S and environment, see Chap. 4 for the formal mathematical presentation based on the apparatus of lifting maps (quantum information theory). Since we use the Hilbert space representation, the adaptive dynamics model under consideration can be called quantum(-like).

References

1. Kolmogoroff, A.N.: Grundbegriffe der Wahrscheinlichkeitsrechnung. Springer, Berlin (1933); English translation: Kolmogorov, A.N.: Foundations of Theory of Probability. Chelsea Publishing Company, New York (1956)
2. Boole, G.: An Investigation of the Laws of Thought. Dover Edition, New York (1958)
3. Von Neuman, J.: Mathematical Foundations of Quantum Mechanics. Princeton University Press, Princeton (1955)
4. Khrennikov, A.: Contextual Approach to Quantum Formalism. Springer, Berlin-Heidelberg-New York (2009)
5. Dirac, P.A.M.: The Principles of Quantum Mechanics. Clarendon Press, Oxford (1995)
6. Khrennikov, A.: Detection model based on representation of quantum particles by classical random fields: Born's rule and beyond. Found. Phys. **39**, 997–1022 (2009)
7. Ingarden, R.S., Kossakowski, A., Ohya, M.: Information Dynamics and Open Systems: Classical and Quantum Approach. Kluwer, Dordrecht (1997)
8. Ohya, M., Volovich, I.V.: Mathematical Foundations of Quantum Information and Computation and Its Applications to Nano- and Bio-systems. Springer, Berlin-New York (2011)
9. Asano, M., Ohya, M., Khrennikov, A.: Quantum-like model for decision making process in two players game. Found. Phys. **41**(3), 538–548 (2010)
10. Basieva, I., Khrennikov, A., Ohya, M., Yamato, I.: Quantum-like interference effect in gene expression: glucose-lactose destructive interference. Syst. Synth. Biol. **5**, 59–68 (2011)
11. Pothos, E.M., Busemeyer, J.R.: A quantum probability explanation for violation of rational decision theory. Proc. R. Soc. B **276**(1165), 2171–2178 (2009)
12. Wang, Z., Busemeyer, J. R.: A quantum question order model supported by empirical tests of an a priori and precise prediction. Top. Cogn. Sci. (2013), to be published

13. Zorn, C.H., Smith, Ch.: Pseudo-classical nonseparability and mass politics in two-party systems. In: Song, D., Melucci, M., Frommholz, I., Zhang, P., Wang, L., Arafat, S. (eds.) Quantum Interaction-3. Lectures Notes in Computer Science, vol. 7052, pp. 83–94. Springer, Heidelberg (2011)
14. Khrennikova, P., Khrennikov, A., Haven, E.: The quantum-like description of the dynamics of party governance in the US political system (2012). arXiv:1206.2888 [physics.gen-ph]
15. Einstein, A., Podolsky, B., Rosen, N.: Can quantum-mechanical description of physical reality be considered complete? Phys. Rev. **47**(10), 777–780 (1933)
16. Fuchs, C.A.: Quantum mechanics as quantum information (and only a little more). In: Khrennikov, A. (ed.) Quantum Theory: Reconsideration of Foundations. Series Mathematical Modeling 2, pp. 463–543. Växjö University Press, Växjö (2002)
17. Caves, C.M., Fuchs, Ch.A., Schack, R.: Quantum probabilities as Bayesian probabilities. Phys. Rev. A **65**, 022305 (2002)
18. Fuchs, Ch.A., Schack, R.: A quantum-Bayesian route to quantum-state space. Found. Phys. **41**, 345–356 (2011)
19. Allahverdyan, A.E., Balian, R., Nieuwenhuizen, Th.M.: The quantum measurement process in an exactly solvable model. In: Foundations of Probability and Physics-3. American Institute of Physics. Series Conference Proceedings, vol. 750, pp. 16–24. Melville (2005)
20. Khrennikov, A.: Description of composite quantum systems by means of classical random fields. Found. Phys. **40**, 1051–1064 (2010)
21. Khrennikov, A., Ohya, M., Watanabe, N.: Classical signal model for quantum channels. J. Russ. Laser Res. **31**, 462–468 (2010)
22. Khrennikov, A.: Ubiquitous Quantum Structure: From Psychology to Finance. Springer, Heidelberg-Berlin-New York (2010)
23. Khrennikov, A.: Prequantum classical statistical field theory: Schrödinger dynamics of entangled systems as a classical stochastic process. Found. Phys. **41**, 317–329 (2011)
24. Khrennikov, A., Ohya, M., Watanabe, N.: Quantum probability from classical signal theory. Int. J. Quantum Inf. **9**, 281–292 (2011)
25. Ohya, M.: Complexities and their applications to characterization of chaos. Int. J. Theor. Phys. **37**, 495–505 (1998)
26. Ohya, M.: Some aspects of quantum information theory and their applications to irreversible processes. Rep. Math. Phys. **27**, 19–47 (1989)
27. Ohya, M., Volovich, I.V.: New quantum algorithm for studying NP-complete problems. Rep. Math. Phys. **52**, 25–33 (2003)
28. Ohya, M.: Adaptive dynamics and its applications to chaos and NPC problem. In: QP-PQ: Quantum Probability and White Noise Analysis. Quantum Bio-Informatics, vol. 21, pp. 181–216 (2007)

Chapter 3
Fundamentals of Molecular Biology

Abstract In this chapter, we briefly introduce the basic concept of molecular biology and also biological information processing in general. As a detailed example of the information processing phenomena (signal recognition, transformation, and biological response) in a living system, we explain the diauxie (two phase) growth of *Escherichia coli* in glucose and lactose mixed medium and show the mechanistic simulation of the system according to the systems biology approach. In addition, we introduce epigenetic mutation as another example for the information processing in a living system.

Keywords Molecular biology · Bio-informational processing · Living systems · Metabolism of *Escherichia coli* · Genetic and epigenetic mutation · Cell differentiation · Physiological contextuality

3.1 Research Fields in Life Science and Information Biology

Life is basically created on the flows of materials, energy, and information. In other words, living organisms "live" in these flows at non-equilibrium steady state. Accordingly, biological research is mainly targeted at the mechanisms of these major flows (Fig. 3.1; [1]). We cite the introduction part from our publication [1] as below and add discussion on our proposal to use "information biology" for the research field of information flow.

During the World War II, a production method of (radio)-isotopes by nuclear reactions was established. Biochemists used such isotopes as tracers to identify many metabolic pathways, such as Calvin cycle [2] in photosynthesis in plants and Krebs cycle [3]. Adenosine triphosphate (ATP) has been found by Fiske and Subbarow [4] as a high energy substance in biochemical reactions. Muscle contraction occurs depending on the consumption of the ATP hydrolysis energy [5], and Huxley and Hanson proposed a sliding model of actomyosin fibers [6]. In muscles, the chemical energy of ATP is converted to mechanical energy to exert force. There are many kinds of energy transducing systems, such as photosynthesis, ion-pumping ATPases, and so on. Then after the proposal of DNA structure by Watson and Crick [7],

M. Asano et al., *Quantum Adaptivity in Biology: From Genetics to Cognition*,
DOI 10.1007/978-94-017-9819-8_3

Fig. 3.1 Research targets of life science (3 major flows)

3 kinds of flows		*"Research field" (examples of researches)*
Flow of materials	→	"biochemistry" (metabolism, etc)
Flow of energy	→	"biophysics" (muscle contraction, etc)
Flow of information	→	"molecular biology" or "information biology" (genetic code, signal transduction, nerve/brain system, etc)

information flows and signal transduction in many biological systems have been extensively elucidated, including highly complex information processing systems as cognition and brain, which are significantly accelerated by the recent development of large-scale and high-throughput research in biology, so called "-omic" studies, such as genome, metabolome, transcriptome, proteome, and interactome, etc.

According to the DNA structure [7], genetic inheritance and information flows from genes to proteins (genetic coding problem) have been studied, which is called molecular biology in a narrow sense dealing with the information processing relating to genes. This information flow is now basically understood as a central dogma in molecular biology [8]. Whereas signal transduction, which deals with the cellular information processing of environmental signals to cellular responses has been studied biochemically; nerve/brain functions have been studied physiologically and biochemically. They are now biochemically and biophysically studied at the molecular level, i.e., molecular biologically. Thus the above named three research fields cover mostly overlapping biological phenomena and the naming of them is rather confusing and arbitrary. In Fig. 3.1, we proposed a classification of research fields according to the kinds of research targets and also to the historical origin as explained above. We included biochemical and biophysical research on information processing (signal transduction and nerve/brain system, etc.) into *molecular biology* in a wide sense, as dealing with any biological information processing. Or the naming of molecular biology may mean any kind of biological research at the molecular level in general, therefore may be taken to include both biochemistry and biophysics. If we take this standpoint, a new name for the research field dealing with any information processing in biology, such as "information biology" as proposed in a former (or in the preface) chapter of this book, will be necessary: "Bioinformatics" seems to be appropriate but many biologists may limit it as the research or analysis on any kind of static biological data, such as DNA and amino acid sequences, with recent development of simulation studies based on the structure biology data. Thus this term may not be adequate for the general naming of researches on any biological information processing. Therefore, we proposed the use of *molecular biology* for the research on information processing in biology, by expanding the research fields not only on the genetic coding problem, but also on any information processing in biology (Fig. 3.1). Furthermore, in this book, we prefer using the *information biology* as a new terminology for this research field because in the latter chapters we shall show that our quantum-like formulation of any information processing in biology gives the first holistic view of the information biology.

At the first conference of the quantum-bio-informatics research center (QBIC) held in Japan in 2007, we presented basics of *in silico* biology to reproduce living systems in a computer based on the recent -omic data [9]. We expected to obtain mechanistic parameters of whole biochemical reactions in a cell and, finally, to simulate the living system in a computer, which is the systems biology approach. In this chapter, we introduce basics of classical molecular biology and information processing in living organisms. As an example of the information processing, we present diauxie effect of *Escherichia coli* (*E. coli*) growth explaining its molecular biological basis and show an approach according to the systems biology for the analysis of diauxie. In addition, we introduce epigenetic mutation as another example of information processing.

3.2 Molecular Biology and Genome

At the era of the birth of the Earth, meteorites are thought to have fallen down from universe with amino acids and bases in them, which are basic units of proteins and nucleic acids, respectively. Somehow they formed ribonucleic acid (RNA), the process of which is not well known. RNA is composed of ribose, bases, and phosphate (Fig. 3.2). The existence of OH group at $2'$ position of ribose is important, that makes RNA reactive. RNA can cut/form covalent bonds of itself or peptide bonds. Furthermore, it can form double helix by the hydrogen bonds with bases in another RNA strand having complementary base sequences, which enables the information of the base sequence to be conserved and transferred to the other RNA. In other words, this enables RNA to multiply with conserved base sequence. Thus RNA is a kind of living things in the sense that it has the ability of assimilation and information transmission to the next generation. The information transfer process can be perturbed. This can generate mistakes in the base sequence conservation by accidents. This is thought to be the first step of evolution (chemical evolution) of living systems. Thus, at the beginning of the Earth and evolution of living things, the RNA world [10] should have been prevailing.

Then RNA somehow created DNA with a similar property but without high reactivity of RNA. They were thought to be covered with lipid bilayers (biomembranes) to give birth to the first living organism, a primitive prokaryote.

In 1953, Watson and Crick reported a double helix model of DNA structure [7]. DNA is now known to be responsible for genetic information transfer to the next generation. As described above, the information of genes can be transmitted to daughter cells by the conserved base sequences determined by hydrogen bonding complementarity. After this discovery of DNA structure, the central dogma in molecular biology was proposed, which suggested that the information of DNA sequence as a gene is transcribed to a messenger RNA (mRNA) and it is translated into a protein (Fig. 3.3) [8]. Proteins are major players in living systems. They work as enzymes to catalyze chemical reactions or constituents in living bodies and create living systems with other substances. Proteins are polymers consisting of amino acids, whose sequences are determined by the base sequences with coding rule of triplet (Table 3.1). DNA (or RNA) are composed of 4 kinds of bases, guanine (G),

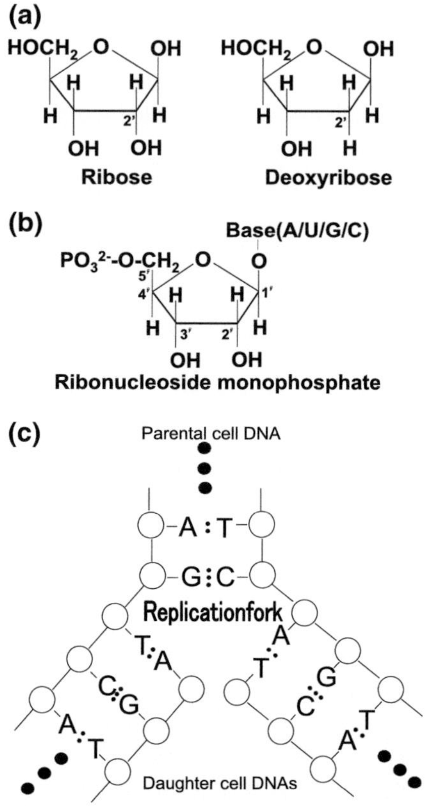

Fig. 3.2 Structures of ribose, deoxyribose, ribonucleotide and synthesizing DNA. **a** Chemical structures of ribose and deoxyribose are shown. The OH at 2′ site in 5 member ring of ribose is important and responsible for high reactivity of RNA. **b** Bases of adenine (A), uridine (U) (thymine (T) in case of DNA), guanine (G), and cytosine (C) are covalently linked to the OH group at the 1′ site of ribose or deoxyribose. They are called ribonucleoside or deoxyribonucleoside. Phosphate is linked to the OH at 5′ site forming a ribonucleoside monophosphate. This phosphate forms bridge with the 3′ OH of the next nucleotide forming RNA in case of ribonucleotides and DNA in case of deoxyribonucleotides. **c** Bases in DNA or RNA can form hydrogen bonding (represented by 2 dots or 3 dots between adjacent two bases on different strands depending on the hydrogen bond numbers) with complementary bases (A:T(U) and G:C) which enables the long DNA or RNA sequences to form double helix with complementary strands, which is the basic mechanism of genetic information conservation into next generation. In this specific figure, replicating DNA is drawn schematically. The deoxyribose rings are shown as *circles* and the phosphate ester bridges are represented by *bars* connecting the *circles* (deoxyribose), resulting in a long chain of nucleotides, which are called as DNA (in case of deoxyribose). The replication apparatus unwinds the helix of parental DNA (*upper part*) and synthesizes new daughter DNA strands complementary to the parental DNA strands at the replication fork, which is the mechanism of genetic information conservation

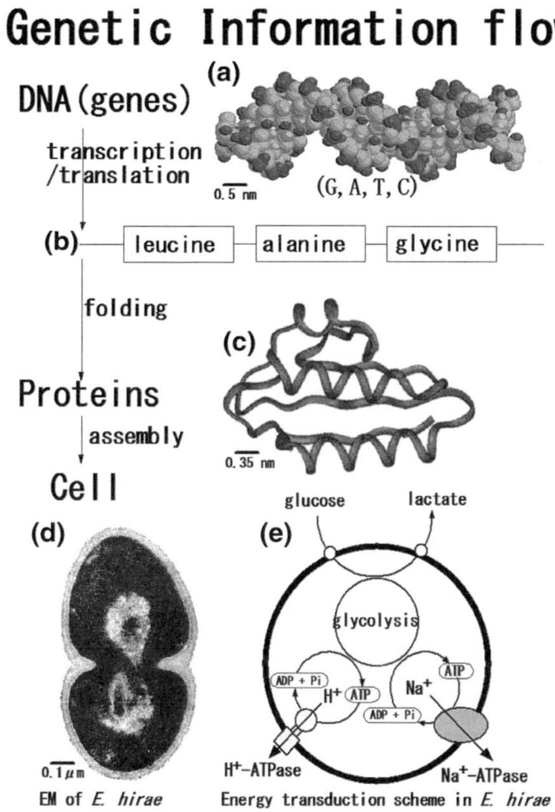

Fig. 3.3 Genetic information flow from DNA (RNA) to proteins to living systems [1]. The base sequences on DNA (**a**) are transcribed into the sequences on RNA, which are translated into amino acid sequences in proteins (**b**). Proteins fold by themselves to form well-defined three dimensional structures (**c**). The assembly of proteins and other components create a living system, such as an *Enterococcus hirae* (*E. hirae*) prokaryotic cell (**d**), having a representative metabolic system shown in (**e**). **c** A schematic representation of the three dimensional structure of a protein with helices and sheets. **d** Electronmicrograph of a bacterium, *E. hirae*. **e** A scheme of the metabolic pathway in *E. hirae*, showing uptake of glucose outside, glycolytic digestion of glucose producing lactate which is exported outside with production of ATP, which then is utilized to pump out inside H^+ and Na^+ by membrane bound ion-translocating ATPases

adenine (A), thymine (T) (uridine, U), and cytosine (C), which are responsible for the information transfer by their hydrogen bonding complementarity (see Fig. 3.2). The combination of 3 bases in sequence makes up one word, thus, in principle, there can be 64 (=4 × 4 × 4) different words, where each word corresponds to a certain amino acid in proteins. However, proteins are composed of only 20 kinds of amino acids. Therefore, there are several redundancies in the usage of triplets (the table, such as Table 3.1, is called codon usage table). Proteins fold into the well-defined three dimensional structures according to their amino acid sequences in a

Table 3.1 Codon usage table

1st nucleotide	2nd nucleotide				3rd nucleotide
	U	C	A	G	
U	Phe	Ser	Tyr	Cys	U
	Phe	Ser	Tyr	Cys	C
	Leu	Ser	Term	Term	A
	Leu	Ser	Term	Trp	G
C	Leu	Pro	His	Arg	U
	Leu	Pro	His	Arg	C
	Leu	Pro	Gln	Arg	A
	Leu	Pro	Gln	Arg	G
A	Ile	Thr	Asn	Ser	U
	Ile	Thr	Asn	Ser	C
	Ile	Thr	Lys	Arg	A
	Met	Thr	Lys	Arg	G
G	Val	Ala	Asp	Gly	U
	Val	Ala	Asp	Gly	C
	Val	Ala	Glu	Gly	A
	Val	Ala	Glu	Gly	G

This shows the corresponding list of triplets of base sequences versus amino acids. The combination of the 1st, 2nd, and 3rd nucleotides in the list makes up a triplet codon to be translated into an amino acid. There are 4 kinds of nucleotides, and triplets can code $4 \times 4 \times 4 = 64$ different kinds of amino acids. There are redundancies, and the triplets represent 20 amino acids as shown in the table. Term means the termination codon signaling the termination of the protein synthesis process

self-organization manner [11]. Atomistic structures of proteins have been solved by a X-ray crystallography and a nuclear magnetic resonance technique.

The genetic information coded by the base sequences in DNA is thus transferred into proteins and finally into living systems of organisms. The whole set of the genes in one organism is called "genome". Therefore, the genome information is essential and may be enough to create a life [12].

3.2.1 Protein Folding Problem

As we have described above, a protein spontaneously folds into its certain three dimensional structure determined by its amino acid sequence. The folding process of proteins typically takes place on the temporal scale from millisecond to second. The folding involves many kinds of interactions between atoms in proteins as well as surrounding solvent molecules/atoms, such as van der Waals, electrostatic, hydrogen bonding, and hydrophobic interaction in water. A protein with 100 amino acids, for example, may consist of about 2,000 atoms, which are surrounded by many water molecules. In 1969, Levinthal showed a paradox in a theory of protein folding [13]:

Let us assume that proteins search their most stable structures by checking the stability one by one through changing every dihedral angle in the protein backbone, in which each amino acid residue has been assumed to take only 9 stable conformations in dihedral angles. For a 100 amino acid protein (since in chemistry transition time of one dihedral angle is about 100 ps), the search would take $10^{-10}\,\mathrm{s} \times (9)^{100}$ dihedral bonds, corresponding to $\sim 8 \times 10^{77}$ years, which is far beyond the time of universe! Therefore, he proposed that some secret mechanisms should be adopted in the protein folding process. The protein folding mechanism has been suggested to follow a glass transition theory in statistical mechanics by Bryngelson et al. as a funnel model [14, 15]. Currently, the whole processes of the folding of several small proteins have been successfully simulated by an atomic level molecular dynamics simulation using a specially designed CPU, ANTON [16].

In this book, we formulate an adaptive dynamics for phenomenological descriptions of biological events as a whole. We think that it is important for each composite element to have interactions with other members forming an interaction network. Even in the protein folding process, each composite atom is interacting with other members by various kinds of forces. Accordingly, we can imagine that to reach its final structure, the folding process must be as quick as a quantum computer: Each step cannot be described according to classical total probability conservation law. Therefore, this system may be a good example of such adaptive dynamics. An example of such system has been described in our previous publication [17].

3.3 Various Information Transductions in Biology

As described in Fig. 3.1, there are many kinds of information transducing processes in biology. Above we described the central dogma underlying the genetic coding mechanism. In biology, we study many other kinds of information transducing processes, such as signal transduction, cognition (brain system), evolution, ecological system, etc. Signal transduction plays major roles in many biological phenomena such as adaptation, reception/response of environmental signals such as hormones, development, differentiation, embryogenesis, etc. A representative example is lactose/glucose diauxie phenomena of *E. coli* growth. It contains several information processing steps, such as signal recognition, interference by signals of other cellular processes, and gene expression regulations. The lactose operon is the first gene expression regulatory system described by Jacob and Monod [18], who were the Nobel Laureates for the discovery and proposal (operon theory). We explain in detail the lactose operon and the diauxie phenomena of *E. coli* in terms of molecular biology in the next section. Other signal transduction systems, such as differentiation, development, and so on, can be described analogously in the sense that their information processing system is composed of network systems with many elements interacting with each other. Adaptive dynamics is powerful in dealing with such complex systems in a phenomenological way. In further chapters we shall show that every such system violates classical total probability conservation law. Some of the examples have been presented previously [17, 19].

Other biological systems of brain/cognition, evolution, ecological systems, etc., even including economical activities, can be similarly formulated by the quantum-like formalism. One of the authors of this book recognized that the total probability conservation law was violated in biological phenomena, such as cognition, psychology, and so on, from the viewpoint of the analogy to quantum mechanical formalism [20]. Another author invented and formulated adaptive dynamics during his study on quantum information theory by taking into account that the observation or interaction with the environment is essential in the real world to give the quantum mechanical behavior in physics [21]. By our collaboration [17, 19], we found that this adaptive dynamics is powerful to describe and formulate the quantum-like features in biological phenomena. Such topics on the concept and tools of quantum-like formulation of biological phenomena are discussed in the latter chapters of this book. They have many composite elements interacting with each other, forming networks, thus one element influences every other element and vice versa. Furthermore, they interact at the same time, summing up all effects at instance. This is the same working principle as quantum computer, and the response should be very quick and already optimized. Our adaptive dynamics can lead to phenomenological formulation of these phenomena, which is the main proposal of our book.

3.3.1 Molecular Biology of Diauxie of E. coli *(Glucose Effect)*

This section is based on the publications [22–25]. Lactose/glucose diauxie phenomenon (biphasic growth) of *E. coli* is described in Fig. 3.4. Lactose operon is composed of three adjacent genes, *lacZ*, *lacY*, and *lacA* (genes are italicized). *lacZ* encodes β-galactosidase, which hydrolyzes lactose into glucose and galactose. *lacY* encodes lactose permease, which is responsible for the active transport of lactose into a cell. *lacA* encodes acetyltransferase. The expressions of these genes are minor in the usual culture media. However, once lactose is present in the culture media, the expressions of the genes are induced to maximum level (Fig. 3.4b). For regulating the gene expression, there is another protein member, lactose repressor coded by *lacI* gene, which binds to the operator part (O) of the operon regulatory region to repress the expression of the lactose operon (Fig. 3.4b). In the presence of lactose, a metabolite, allo-lactose, is produced in the cell and binds to the LacI to induce conformational change of the protein, resulting in unbinding of the LacI from the operator region. Then the RNA polymerase can bind to the promoter region (P) in the lactose operon regulatory region and start transcription to form mRNA of the lactose utilization genes. The mRNA is used for the translation to produce proteins, LacZ and LacY, which work for lactose utilization of the cell. If CRP (cAMP receptor protein) or CAP (catabolite gene activator protein) binds cAMP (cyclic AMP), the cAMP-CRP complex binds to CRP site locating upstream of the *lac* promoter and assists binding of the RNA polymerase resulting in enhancement of gene transcription. cAMP

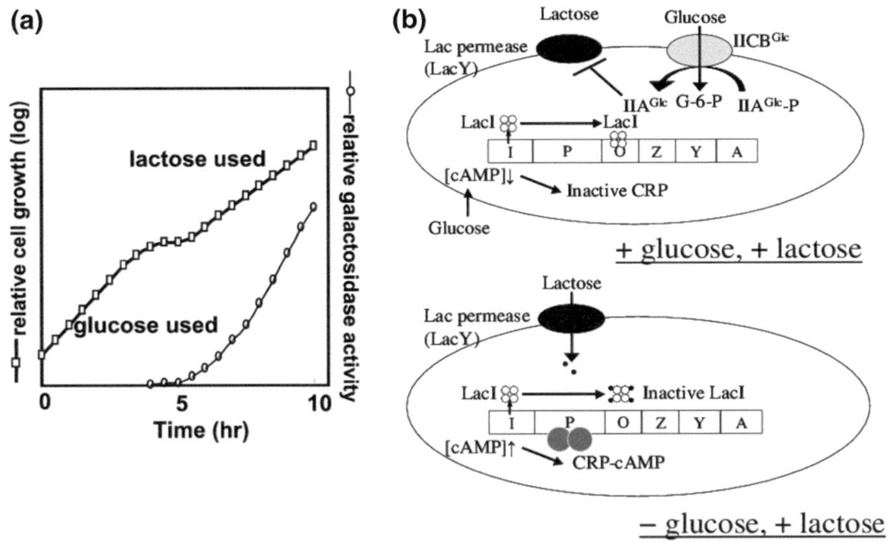

Fig. 3.4 *E. coli* diauxie phenomenon. **a** The growth of *E. coli* was monitored in time dependent manner. The cells grown in the presence of low concentration of glucose and usual concentration of lactose shows biphasic growth character. This classical observation has been known for long [24] and the figure was drawn schematically, showing the biphasic growth curve and the expression of β-galactosidase upon consumption of glucose in the medium. **b** The schematic molecular mechanism of the diauxie [19]. The details are described in the text. *Big ovals* represent cells. *Boxes* with I, P, O, Z, Y and A inside represent genes of lactose repressor and lactose operon on *E. coli* genome DNA. *Black ovals* on the cell membrane represent lactose permease (LacY) responsible for the lactose transport into cells. *Grey ovals* represent EnzymeIICB^{Glc} in glucose phosphotransferase system (PTS) working for transport and phosphorylation of glucose outside forming glucose-6-P inside cells

is produced from ATP by adenylate cyclase in cell membrane when the protein is activated by EnzymeIIA^{Glc} (Figs. 3.4 and 3.5).

The mechanism of the diauxie is explained by catabolite repression and inducer exclusion [22, 23]. Catabolite repression means repression of the gene expression. Glucose is taken up by *E. coli* through glucose phosphotransferase system (PTS) (Fig. 3.5; [26]). Phosphoenolpyruvate inside cells is used to form phospholyrated HPr (*h*eat stable *pr*otein or *h*istidine containing *pr*otein), HPr-P, by EnzymeI^{Glc}. Next, phosphate in HPr-P is transferred to EnzymeIIA^{Glc} to form the phosphorylated form of EnzymeIIA. The phosphate in EnzymeIIA is used to phosphorylate and transport outside glucose to make inside glucose-6-P by EnzymeIICB^{Glc} in membrane. The phosphorylated form of EnzymeIIA^{Glc}, which is produced and maintained in large amount in the absence of glucose in medium, activates the membrane bound adenylate cyclase to produce large amount of cAMP, thus advantageous for lactose operon gene expression. The non-phosphorylated EnzymeIIA^{Glc}, which is produced in large amount when glucose outside is transported, inhibits the LacY, lactose permease in membrane. As a consequence of the inhibition, lactose (inducer) cannot be transported into the cell. Therefore, this mechanism is called inducer exclusion.

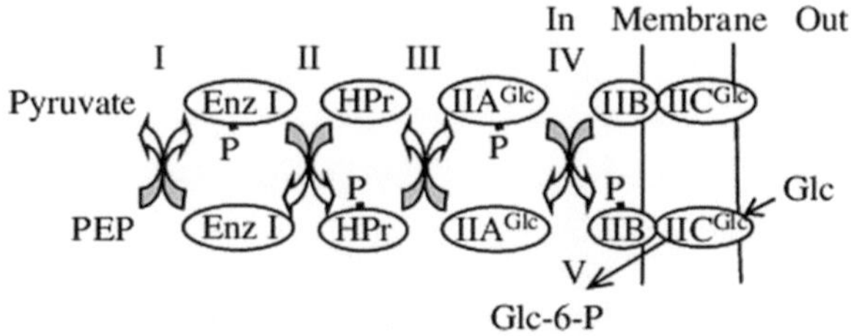

Fig. 3.5 Glucose phosphotranspherase system (PTS) in *E. coli* [26]. The details are described in the text. The first reaction (*I*) in the PTS system is the phosphate transfer from phosphoenolpyruvate (PEP) to His residue in the Enzyme I producing pyruvate. The second transfer process of phosphate (*II*) is the phosphorylation of His residue in HPr (His containing or *h*eat stable *pr*otein) by Enzyme I phosphate. The third reaction (*III*) is the phosphate transfer from HPr-P to Enzyme IIA^{Glc} (or Crr = *c*atabolite *r*epression *r*esistant), the phosphate of which in turn is transferred (*IV*) to Enzyme IIB in membrane bound Enzyme IICB (PtsG) complex, which then is transferred to glucose outside transported through Enzyme IICB complex forming glucose-6-phosphate inside the cell (*V*). IIA is the key component in the regulation of diauxie. The phosphorylated form of IIA is the signal of the absence of glucose outside because without glucose transport, the phosphate group in IIA can be kept for a long time without consumption to form Glc-6-P, thus phosphorylated form of IIA activates membrane bound adenylate cyclase, which produces cAMP, a signal substance of catabolite repression release. cAMP in turn is bound to CRP or CAP to form complex, which binds to the promoter region of lactose operon DNA, which then stimulates the RNA polymerase binding to the promoter region of the lactose operon, thus activating the lactose operon gene expression. On the contrary, the non-phosphorylated form of IIA is the signal of the presence of glucose outside because during glucose transport the phosphate group in IIA is constantly consumed to be transferred to glucose, thus IIA non-phosphorylated form accumulates and directly inhibits the lactose permease activity inhibiting lactose transport, which is called inducer exclusion

As a result, *E. coli* establishes its system to utilize glucose more preferentially than lactose, which is convenient for the cell, because glucose is the easier starting substrate for energy metabolism in glycolysis pathway than lactose. (In order to be utilized, lactose should be first hydrolyzed to galactose and glucose by β-galactosidase.)

Summing up the aforesaid (Fig. 3.4), when glucose and lactose are present in medium, *E. coli* cells tend to utilize glucose first by inhibiting lactose utilization by gene expression repression owing to the low cAMP (catabolite repression) and LacY inhibition (inducer exclusion). After glucose consumption, cells recognize that there is no glucose in the medium and start to change the cellular system to be able to utilize lactose, which is the adaptation to environment by raising cAMP concentration to start gene expression and by stopping inducer exclusion to take up the inducer, lactose. During this changing period of the cellular system, cells start to produce β-galactosidase and stop growth, which brings about the two step growth phenomenon.

As understood by the above description, this process includes the reception of a signal outside, signal transduction through cAMP production and LacY inhibition, and regulation of the gene expression. These kinds of players similarly play important roles in the development, differentiation, and embryogenesis. Thus, their features should be similarly described by our adaptive dynamics.

Cognition (brain system), evolution, and ecological systems have similar mutual network interactions between member elements. Therefore, we can expect that they can be described essentially in a similar manner. We would like to describe our systems biology approach to the diauxie phenomenon in the next section to compare it with our theoretical, phenomenological, and holistic approach described in the latter chapters.

In this sense, the adaptive dynamics is very useful to describe complex systems from micro (atomic level interaction leading to protein self-organization, folding) to macro (species or individual interaction leading to ecological state on Earth). The "adaptive" is a good term because it is suitable to describe such events as adaptation in biology: This adaptation is also the main reaction in the differentiation, development, and embryogenesis. We hope that the readers of this book realize that the biological systems can be modeled as such quantum-like systems finding out optimized responses (answers) very quickly and efficiently.

3.3.2 Systems Biological Approach to the Diauxie (Computer Simulation)

In the previous section, we described basic molecular mechanism of the diauxie of *E. coli*. Since -omic studies become popular, the systems biology has been developed quickly and there are many system simulation tools reported [27–29].

Such an approach to *E. coli* diauxie has been reported [30]. We also developed a system simulation tool (cell simulator) written in C++ (manuscript in preparation) on the basis of biochemical data and an enzymological reaction theory, such as a Michaelis-Menten kinetics. We used parameters obtained from experimental data which we could find in order to directly compare the simulation and experimental results. Here we describe the partial results of our trials in order to make clear the difference between these approaches; we have explained the mechanism elucidated by the molecular biological (experimental) approach in the above section, and in this section we describe the computer simulation (systems biology) approach. In the latter chapters of this book, the theoretical approach for quantum information biology shall be presented.

We performed the cellular simulation of metabolic enzymes and gene expression/regulation activities in *E. coli* growing in the presence of glucose and lactose. The system was composed of the gene regulation system (lactose operon, *lacI* gene, *crp* gene), metabolic pathway (glycolytic and lactose-utilizing pathways), and phosphotransferase system (Fig. 3.6a). We also took into account the inducer exclusion mechanism (inhibition of lactose permease) and the glucose effect (reducing intracellular cAMP level upon glucose transportation into cells). We simulated this system and analyzed the results showing that the ATP production (taken as the cell growth) was biphasic (Fig. 3.6b), the first phase corresponding to the glucose consuming phase and the second to the lactose consuming phase with the intermittent growth ceasing phase when the cells started production of β-galactosidase.

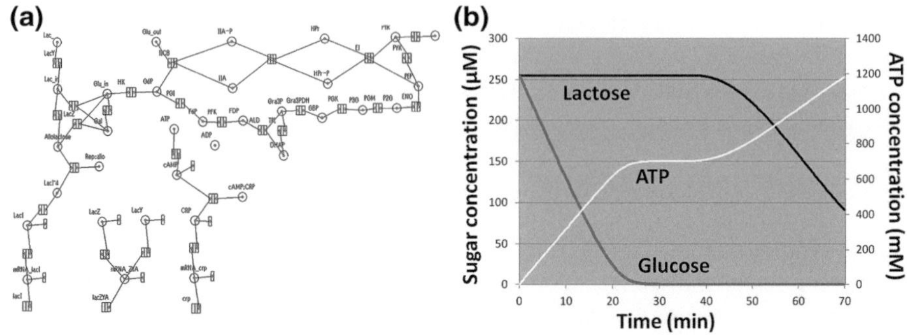

Fig. 3.6 System simulation of the diauxie growth of *E. coli*. **a** Schematic presentation of the system we used as shown on a PC monitor. This system consisted of glycolytic network, gene expression/regulation, and signal transduction (inducer exclusion and catabolite repression) as described in the text. *Circles* represent proteins and metabolites, *boxes* represent reactions, and *edges* represent the connection between proteins, metabolites, and their reactions. **b** Simulation result showing the ATP concentration in the cell during growth in the presence of glucose and lactose mixture. ATP increased biphasically mimicking the biphasic growth of cell, diauxie. During the first growth phase, glucose in the medium was consumed and lactose retained. During the second growth phase, after a short period of cease of growth, lactose was consumed. The β-galactosidase production (*lacZ* gene expression) was induced during the intermittent growth stop phase after the consumption of glucose and the products accumulated to be used for lactose hydrolysis and utilization

3.3.3 Epigenetic Mutation and Evolution

Our living organisms utilize three typical regulation mechanisms for regulating flows of matter (metabolism etc.), energy, and information (such as adaptation that we described above, etc.). The first one is the allosteric effect or ligand induced conformational change causing functional change of a protein, an example of which in the information flow is the allo-lactose induced LacI repressor dissociation from the operator region of lactose regulatory region or the inducer exclusion caused by the non-phosphorylated EnzymeIIA^{Glc} inhibiting directly LacY permease. The second one is the chemical modification of a protein changing its functional activity, an example ofwhich in the information flow is the phosphorylation or de-phosphorylation of EnzymeIIA^{Glc} changing its activity, on the one hand activating adenylate cyclase producing cAMP for gene expression stimulation and on the other hand inhibiting the LacY permease directly bringing about the inducer exclusion. The above two mechanisms are suitable for quick response to an environmental change (in a few seconds or in a few minutes). The third one involves the gene expression regulation leading to new protein production, which usually takes more time (a few hours) but still within the individual own generation, an example of which in the information flow is the gene expression regulation of lactose operon depending on the presence of lactose or glucose. We described these mechanisms in detail above (Sect. 3.3.1 and Fig. 3.4). But these three mechanisms do not utilize mutations (base sequence changes in genome) in any sense.

In the case of diauxie (Sect. 3.3.1), the adaptation process is not accompanied by any base sequence changes (mutation) or modification of the DNA. Still, the adapted phenotype of the parental cell can be transmitted to the daughter cells for several generations even in the absence of the specific environmental pressure, which is an apparent indication of epigenetic mutation. But for such single cell organisms, we do not call this phenomenon epigenetic mutation, and we rather classify as adaptation.

In multicellular organisms, adaptation takes place in their bodies (usually in somatic cells) by mechanisms with or without modifications of their chromatin structures. The phenotypic change caused by the chromatin structure modifications without base sequence changes is usually called epigenetics. The mechanism seems quite similar to that adopted in the above adaptation process in *E. coli* (Sect. 3.3.1); the environmental pressure stimulates or inhibits the chromatin modification system to induce or repress the relevant gene expression. The only difference is that epigenetics utilize chromatin structure modification such as DNA methylation or histone acetylation, etc., without base sequence changes. Recently it has been known that such epigenetic effect can be transmitted to the next generation called epigenetic mutation or epimutation in order to distinguish it from the usual mutation caused by base sequence changes. (As understood from the above explanation, the definitions of epigenetics and epimutation are not unequivocal.)

The epigenetics is a kind of adaptation and the mechanism of the appearance of epigenetics should be similar to the adaptation processes. But in the case of epimutation, the phenotype is transmitted to the next generation even in the absence of the environmental pressure. Therefore, epimutation is thought to be the underlying mechanism leading to a Lamarckian evolution. Thus it is easy to understand that such mutation and evolution are fast as compared with the usual evolution taken place by the Darwinian mechanism with random mutations and selection under certain environmental pressure. We formulated such epimutation and evolution processes using our quantum-like formalism in the latter chapters [31].

However the final answer or response obtained by epimutations or random mutations/selection processes should be the same since both evolution processes occur under the condition with every composite element or member interacting with each other and also with the environment forming a big interacting network similar to the system of adaptation. The only difference is the speed of the appearance. In the latter chapters of this book, we present the application of our quantum-like formalism or adaptive dynamics to the epimutation phenomena.

References

1. Yamato, I.: From structure and function of proteins toward *in silico* biology. Quantum Bio-Informatics V. (From Quantum Information to Bio-Informatics). In: Accardi, L., Freudenberg, W., Ohya, M. (eds.) Proceedings of the International Symposium of Quantum Bio-Informatics Research Center 2011, Chiba, pp. 473–485. World Scientific, Singapore (2013)
2. Bassham, J., Benson, A., Calvin, M.: The path of carbon in photosynthesis. J. Biol. Chem. **185**, 781–787 (1950)

3. Krebs, H.A., Johnson, W.A.: Metabolism of ketonic acids in animal tissues. Biochem. J. **31**, 645–660 (1937)
4. Fiske, C.H., Subbarow, Y.: Phosphorus compounds of muscle and liver. Science **70**, 381–382 (1929)
5. Szent-Györgyi, A., Banga, I.: Adenosinetriphosphatase. Science **93**, 158 (1941)
6. Huxley, H., Hanson, J.: Changes in the cross-striations of muscle during contraction and stretch and their structural interpretation. Nature **173**, 973–976 (1954)
7. Watson, J.D., Crick, F.H.C.: Molecular structure of nucleic acids: a structure for deoxyribose nucleic acid. Nature **171**, 737–738 (1953)
8. Crick, F.: Central dogma of molecular biology. Nature **227**, 561–563 (1970)
9. Yamato, I., Ando, T., Suzuki, A., Harada, K., Itoh, S., Miyazaki, S., Kobayashi, N., Takeda, M.: Toward in silico biology (from sequences to systems). Quantum Bio-Informatics (From Quantum Information to Bio-Informatics). In: Accardi, L., Freudenberg, W., Ohya, M. (eds.) Proceedings of the International Symposium of Quantum Bio-Informatics Research Center 2007, Chiba, pp. 440–455. World Scientific, Singapore (2008)
10. Gilbert, W.: The RNA world. Nature **319**, 618 (1986)
11. Anfinsen, C.B.: Principles that govern the folding of protein chains. Science **181**, 223–230 (1973)
12. Gibson, D.G., Benders, G.A., Andrews-Pfannkoch, C., Denisove, E.A., Baden-Tillson, H., Zaveri, J., Stockwell, T.B., Brownley, A., Thomas, D.W., Algire, M.A., Merryman, C., Young, L., Noskov, V.N., Glass, J.I., Venter, J.C., Hutchison 3rd, C.A., Smith, H.O.: Complete chemical synthesis, assembly, and cloning of a Mycoplasma genitalium genome. Science **319**, 1215–1220 (2010)
13. Levinthal, C.: How to fold graciously. Mossbauer Spectroscopy in Biological Systems: Proceedings of a Meeting Held at Allerton House, pp. 22–24. Monticello, Illinois (1969)
14. Bryngelson, J.D., Wolynes, P.G.: Spin glasses and the statistical mechanics of protein folding. Proc. Natl. Acad. Sci. USA **84**, 7524–7528 (1987)
15. Bryngelson, J.D., Onuchic, J.N., Socci, N.D., Wolynes, P.G.: Funnels, pathways, and the energy landscape of protein folding. Proteins: Struct., Funct., Bioinform. **21**, 167–195 (1995)
16. Shaw, D.E., Maragakis, P., Lindorff-Larsen, K., Piana, S., Dror, R.O., Eastwood, M.P., Bank, J.A., Jumper, J.M., Salmon, J.K., Shan, Y., Wriggers, W.: Atomic-level characterization of the structural dynamics of proteins. Science **15**, 341–346 (2010)
17. Basieva, I., Khrennikov, A., Ohya, M., Yamato, I.: Quantum-like interference effect in gene expression: glucose-lactose destructive interference. Syst. Synth. Biol. **5**, 59–68 (2011)
18. Jacob, F., Monod, J.: Genetic regulatory mechanism in the synthesis of proteins. J. Mol. Biol. **3**, 318–356 (1961)
19. Asano, M., Basieva, I., Khrennikov, A., Ohya, M., Tanaka, Y., Yamato, I.: Quantum-like model of diauxie in *Escherichia coli*: operational description of precultivation effect. J. Theor. Biol. **314**, 130–137 (2012)
20. Khrennikov, A.: Classical and quantum mechanics on information spaces with applications to cognitive, psychological, social and anomalous phenomena. Found. Phys. **29**, 1065–1098 (1999)
21. Ohya, M.: Adaptive dynamics and its applications to chaos and NPC problem. QP-PQ: quantum probability and white noise analysis. Quantum Bio-Inform. **21**, 181–216 (2007)
22. Beckwith, J.R., Zipser, D.: The Lactose Operon. University of Tokyo Press, Tokyo (1971)
23. Beckwith, J.R.: The lactose operon in *Escherichia coli* and *Salmonella typhimurium*. In: Neidhardt, F.C., Ingraham, J.L., Low, K.B., Magasanik, B., Schaechter, M., Umbarger, H.E. (eds.), pp. 1444–1452. ASM, Washington (1987)
24. Loomis Jr., W.F., Magasanik, B.: Glucose-lactose diauxie in *Escherichia coli*. J. Bacteriol. **93**, 1397–1401 (1967)
25. Inada, T., Kimata, K., Aiba, H.: Mechanism responsible for glucose-lactose diauxie in *Escherichia coli* challenge to the cAMP model. Genes Cells **1**, 293–301 (1996)
26. Postma, P.W.: Phosphotransferase system for glucose and other sugars in *Escherichia coli* and *Salmonella typhimurium*. In: Neidhardt, F.C., Ingraham, J.L., Low, K.B., Magasanik, B., Schaechter, M., Umbarger, H.E. (eds.), pp. 127–141. ASM, Washington (1987)

27. Kitano, H.: Systems biology: a brief overview. Science **295**, 1662–1664 (2002)
28. Tomita, M., Hashimoto, K., Takahashi, K., Shimizu, T.S., Matsuzaki, Y., Miyoshi, F., Saito, K., Tanida, S., Yugi, K., Venter, J.C., Hutchison 3rd, C.A.: E-CELL: software environment for whole-cell simulation. Bioinformatics **15**, 72–84 (1999)
29. Mendes, P.: GEPASI: a software for modeling the dynamics, steady states and control of biochemical and other systems. Comput. Appl. Biosci. **9**, 563–571 (1993)
30. Albert, J., Rooman, M.: Dynamic modeling of gene expression in prokaryotes: application to glucose-lactose diauxie in *Escherichia coli*. Syst. Synth. Biol. **5**, 33–43 (2011)
31. Asano, M., Basieva, I., Khrennikov, A., Ohya, M., Yamato, I.: A model of epigenetic evolution based on theory of open quantum systems. Syst. Synth. Biol. (2013). doi:10.1007/s11693-013-9109-2

Chapter 4
Adaptive Dynamics and General Approach to Non-Kolmogorov Probability Theory

Abstract In this chapter, we will discuss the non-Kolmogorov probability theory. Our aim is to explain where and why the usual probability theory is broken. The basic example from quantum physics is presented in very detail including the corresponding illustrations. This is the two slit experiment demonstrating interference for quantum systems. Further, we discuss the mathematical foundation of lifting theory (the basic element of quantum information theory used in this book) and the concept of adaptive dynamics with its mathematical description. Then we apply this mathematical basis to the study of the non-Kolmogorov probability theory. One of the main messages is that, although biological probabilistic behaviors (including cognition) are nonclassical, i.e., it cannot be described by the standard Kolmogorov model, it is not always possible to represent them in the canonical quantum framework. More general quantum-like models have to applied. The last part of this chapter is devoted to quantum informational foundations of theory of adaptive dynamics (AD). Here lifting of quantum states plays the fundamental role. This model generalized the canonical quantum model based on theory of open quantum systems.

Keywords Classical (Kolmogorov) and non-classical (non-Kolmogorov) probabilities · Quantum probability · Two slit experiment · Interference · Lifting · Positive operator valued measures · Adaptive dynamics · Context-dependence

In this chapter, we will discuss the non-Kolmogorov probability theory. Our aim is to explain where and why the usual probability theory is broken. Further, we discuss the mathematical foundation of lifting theory and the concept of adaptive dynamics with its mathematical description. Then we apply these mathematical basis to the study of the non-Kolmogorov probability theory.

M. Asano et al., *Quantum Adaptivity in Biology: From Genetics to Cognition*,
DOI 10.1007/978-94-017-9819-8_4

4.1 Violation of Formula of Total Probability and Non-Kolmogorov Probability Theory

4.1.1 Interference of Probabilistic Patterns in the Two Slit Experiment

The two slit experiment is the basic example demonstrating that QM describes statistical properties in microscopic phenomena, to which the classical probability theory seems to be not applicable, see, e.g., Feynman et al. [1]. In this section, we consider the experiment with the symmetric setting: the source of photons is located symmetrically with respect to two slits, Fig. 4.1. Consider a pair of random variables a and b. We select a as the "slit passing variable," i.e., $a = 0, 1$, see Fig. 4.1, and b as the position on the photo-sensitive plate, see Fig. 4.2. Remark that the b-variable has the continuous range of values, the position x on the photo-sensitive plate. We denote $P(a = i)$ by $P(i)$ $(i = 1, 2)$, and $P(b = x)$ by $P(x)$.

Fig. 4.1 Experimental setup

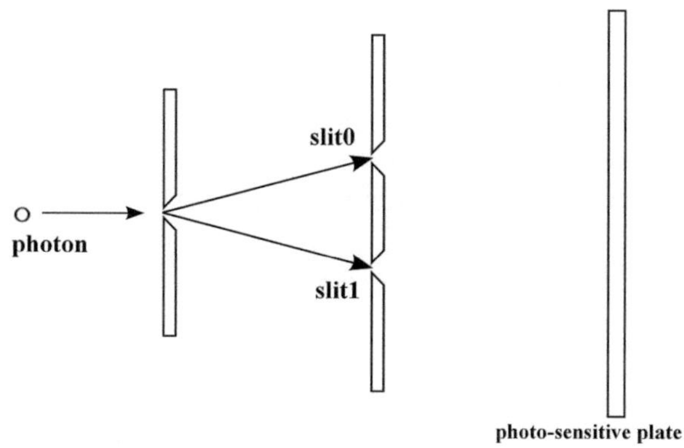

Fig. 4.2 Context with both slits are open

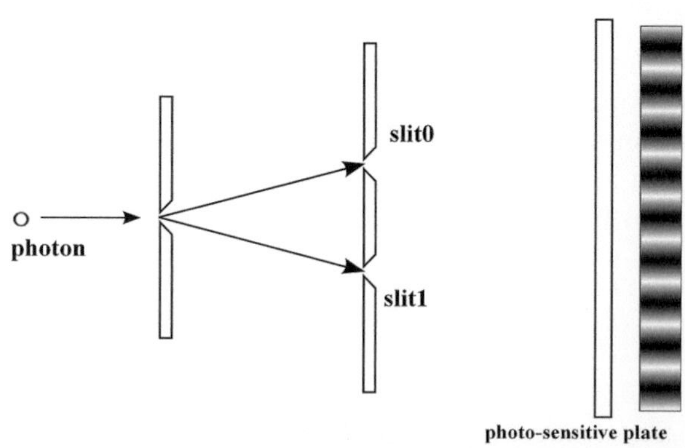

The probability that a photon is detected at position x on the photo-sensitive plate is represented as

$$P(x) = \left| \frac{1}{\sqrt{2}} \psi_0(x) + \frac{1}{\sqrt{2}} \psi_1(x) \right|^2$$
$$= \frac{1}{2} |\psi_0(x)|^2 + \frac{1}{2} |\psi_1(x)|^2 + |\psi_0(x)| \, |\psi_1(x)| \cos\theta, \quad (4.1)$$

where ψ_0 and ψ_1 are two wave functions, whose absolute values $|\psi_i(x)|^2$ give the distributions of photons passing through the slit $i = 0, 1$, see Fig. 4.2.

The term

$$|\psi_0(x)| \, |\psi_1(x)| \cos\theta$$

implies the interference effect of two wave functions. Let us denote $|\psi_i(x)|^2$ by $P(x|i)$, then the Eq. (4.1) is represented as

$$P(x) = P(0)P(x|0) + P(1)P(x|1) + 2\sqrt{P(0)P(x|0)P(1)P(x|1)} \cos\theta. \quad (4.2)$$

Here the values of probabilities $P(0)$ and $P(1)$ are equal to $1/2$ since we consider the symmetric settings. For general experimental settings, $P(0)$ and $P(1)$ can be taken as the arbitrary nonnegative values satisfying $P(0) + P(1) = 1$. In the above form, the classical probability law, the formula of total probability (FTP),

$$P(x) = P(0)P(x|0) + P(1)P(x|1), \quad (4.3)$$

is violated, and the term of interference $2\sqrt{P(x|0)P(0)P(x|1)P(1)} \cos\theta$ specifies the violation.

Typically this violation of FTP is presented as one of quantum mysteries and "the violation of laws of classical probability" is coupled with the exotic features of quantum particles. We do not share this view. In our opinion, the violation of FTP is a consequence of the special contextual structure of the two slit experiment (in fact, a group of experiments).

To find the conditional probabilities $P(x|i)$, experimenters consider two experimental contexts presented on Fig. 4.3, only slit 0 is open, and on Fig. 4.4, only slit 1 is open. The usage of these contexts in attempt to verify FTP is (consciously or even unconsciously) based on the set-theoretical representation of contexts. Suppose that the context of Fig. 4.1 is represented by the set $A = \{(a = 0) \vee (a = 1)\}$ and the contexts of Figs. 4.3 and 4.4 by sets $A_0 = \{a = 0\}$ and $A_1 = \{a = 1\}$. One supposes that $A = A_0 \cup A_1$ and then tries to use FTP with the aforementioned surprise of its violation. In the next section we shall discuss this problem in more detail. Another point is that one has to distinguish (random) variables of classical probability theory from observables. Speaking about variables (as we did at the beginning of this section) there is no need to determine contexts for their measurement (they exist only in the mathematical model). However, speaking about observables one has to specify contexts for measurements.

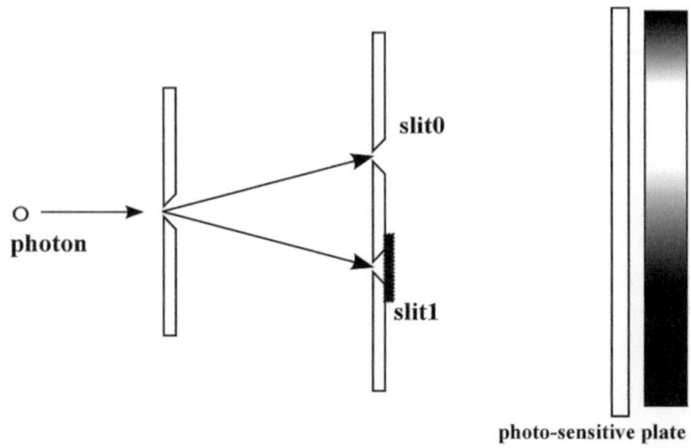

Fig. 4.3 Context with one slit is closed-II

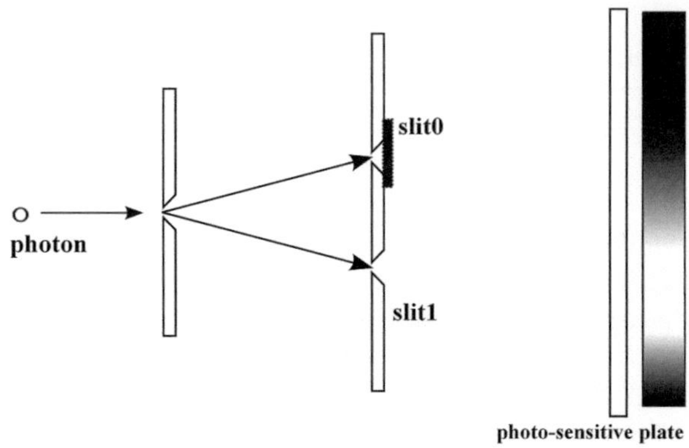

Fig. 4.4 Context with one slit is closed-I

Contextual Viewpoint on Conditional Probabilities and Formula of Total Probability

In this section, we shall repeat the previous considerations on the violation of FTP in the two slit experiment. Now we proceed in the abstract contextual framework, see the monograph [2] for details. We state again that the "two slit experiment" consists of a few experiments with essentially different experimental setups. And, by violating FTP, one operates with statistical data collected in these incompatible experiments. Operating with abstract contexts may be difficult for biologists. Therefore one can read only the beginning of this section devoted to the contextual reformulation of the notion of conditional probability and then jump to formula (4.10).

In classical probability theory we assume that it is possible to form the event $B \cap A$, "both events A and B take place." Suppose that these events correspond to the measurements of two (discrete) random variables a and b. That is, we can consider the sets of the events $A_\alpha = \{a = \alpha\}$ and $B_\beta = \{b = \beta\}$. If these random variables are incompatible, i.e., it is impossible to construct an experimental context for their joint measurement, then the event $C_{\alpha,\beta} = A_\alpha \cap B_\beta$ is meaningless. This happens, e.g., in quantum physics with incompatible observables represented by noncommutative

Fig. 4.5 Context with two detectors

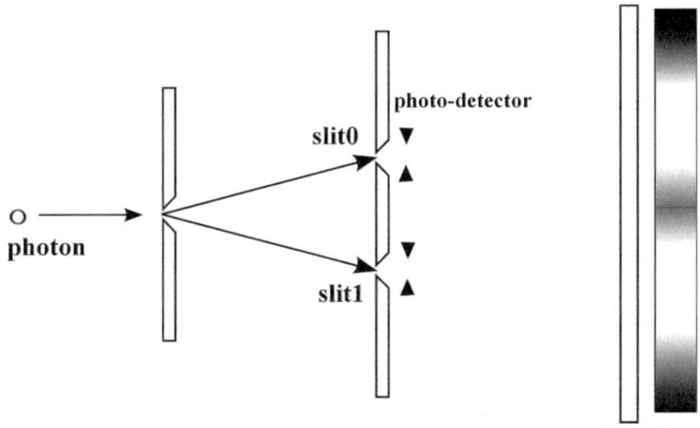

operators. (In Sect. 4.1.2, we discuss joint measurements in a quantum system, and its relation with noncommutativity.) Therefore *conditioning with respect to an event* has a restricted domain of application. In the general situation conditioning has to be treated as conditioning with respect to the *context of measurement* [3].

Hence, in the expression $P(B|C)$ the symbols B and C have to be treated in very different ways.[1] The first one, e.g., $B = B_\beta$, still denotes an event corresponding to the measurement for the value β of the random variable b. However, the symbol C is used to denote context for the measurement of the b-observable. We now turn to FTP. Can we generalize FTP by using context conditioning instead of Kolmogorovian event conditioning?

As was shown, see (4.2), in quantum physics the standard FTP can be violated. To simplify the introduction to the two slit experiment, in Sect. 4.1.1 we studied this experiment in the symmetric setting, the source was located symmetrically with respect to the slits. In such a setting we can use equal probabilities for passing through slits without further explanation. To formalize the general situation, we introduce two (incompatible!) observables: b gives the point x of detection on the registration screen and $a = 0, 1$ gives information about which slit is passed. Both observables are measured under the context C, both slits are open, (Figs. 4.1 and 4.5) and the following contextual probabilities are obtained: $P(b = x|C)$ and $P(a = i|C),\ i = 0, 1$. (In Sect. 4.1.1 these probabilities were denoted as $P(x)$ and $P(i),\ i = 0, 1$). Besides the context C, two other contexts are involved: C_i, only i-th slit is open, $i = 0, 1$, Figs. 4.3 and 4.4. We also measure the b-observable with respect to these two contexts and obtain the contextual probabilities $P(b = x|C_0)$ and $P(b = x|C_1)$. (In Sect. 4.1.1 these were simply $P(x|0), P(x|1)$.) We remark that, for these contexts, we can measure even the a-observable and we have:

$$P(a = i|C_i) = 1, \tag{4.4}$$

---

[1] We started to use the symbol C to denote experimental context; symbols A, B are reserved for events.

$$P(a = i|C_j) = 0, i \neq j. \tag{4.5}$$

As in Sect. 4.1.1, we obtain the generalized FTP as the following formula with the interference term:

$$P(b = x|C) = P(a = 0|C)P(b = x|C_0) + P(a = 1|C)P(b = x|C_1)$$
$$+ 2\cos\theta\sqrt{P(a = 0|C)P(b = x|C_0)P(a = 1|C)P(b = x|C_1)}. \tag{4.6}$$

Thus the classical FTP is violated. As was already pointed in Sect. 4.1.1, the main source of surprise by violation of classical FTP (and, hence, classical probability theory), see, e.g., Feynman et al. [1], is the application of event conditioning and the set-theoretical algebra for events. Motivated by (4.4) and (4.5) people (even experts in quantum foundations) typically identify the contexts $C_i, i = 0, 1$, with the events $A_i = \{a = i\}$ and the context C with the event $A_0 \cup A_1$. One also expects that FTP holds true:

$$P(b = x|A_0 \cup A_1) = P(A_0)P(b = x|A_0) + P(A_1)P(b = x|A_1) \tag{4.7}$$

(with the natural assumptions: $P(A_i) = P(A_i|A_0 \cup A_1)$ and $P(b = x) = P(b = x|A_0 \cup A_1)$).

Now we can repeat the previous contextual analysis in the abstract framework, i.e., without coupling to the two slit experiment. There are two in general incompatible observables, say a and b. For our applications, it is sufficient to consider dichotomous observables, $a = \alpha_1, \alpha_2$ and $b = \beta_1, \beta_2$. There are given the experimental context C and probabilities with respect to it, $P(a = \alpha_i|C)$ and $P(b = \beta_i|C)$. There are also given two other contexts denoted by $C_{\alpha_i}, i = 1, 2$, such that

$$P(a = i|C_{\alpha_i}) = 1, \tag{4.8}$$

$$P(a = i|C_{\alpha_j}) = 0, i \neq j \tag{4.9}$$

cf. (4.4), (4.5). Then [3], for $i = 1, 2$,

$$P(b = \beta_i|C) = P(a = \alpha_1|C)P(b = \beta_i|C_{\alpha_1}) + P(a = \alpha_2|C)P(b = \beta_i|C_{\alpha_2})$$
$$+ 2\lambda_i\sqrt{P(a = \alpha_1|C)P(b = \beta_i|C_{\alpha_1})P(a = \alpha_2|C)P(b = \beta_i|C_{\alpha_2})} \tag{4.10}$$

and the coefficient λ_i, "-E-interference coefficient" , is in general nonzero. This coefficient can be used as a measure of interference between incompatible observables a and b. If it happens (as in the two slit experiment) that

$$|\lambda_i| \leq 1, i = 1, 2, \tag{4.11}$$

we can introduce new parameters, "probabilistic phases", and represent the interference coefficients as

$$\lambda_i = \cos\theta_i. \tag{4.12}$$

Then the generalized FTP (4.10) takes the form:

$$P(b=\beta_i|C) = P(a=\alpha_1|C)P(b=\beta_i|C_{\alpha_1}) + P(a=\alpha_2|C)P(b=\beta_i|C_{\alpha_2})$$

$$+\, 2\cos\theta_i \sqrt{P(a=\alpha_1|C)P(b=\beta_i|C_{\alpha_1})P(a=\alpha_2|C)P(b=\beta_i|C_{\alpha_2})}. \tag{4.13}$$

However, in general *the interference coefficients can exceed one*! This can happen even in quantum theory—for generalized observables a and b, which are represented mathematically not by Hermitian operators, but by *positive operator valued measures* (POVMs), see Sect. 10.1 for details. In quantum bio-informatics it is common that interference is so strong that $|\lambda| > 1$, Sect. 5.1.2, Eq. 5.3. Thus here the standard quantum formalism based on the representation of observables by Hermitian operators is not applicable. One has to use its generalizations, which would describe adequately the biological situation. One possibility is to use POVMs (as people do in quantum information theory). And we test this apparatus in Chap. 10. However, we prefer to apply a novel quantum probabilistic model [4] based on the *entanglement* between a system and an environment. This model is based on such an advanced tool of quantum information theory as a *lifting map*, see Sect. 4.2 for the detailed presentation. This model matches very well with the probabilistic structure of quantum bio-informatics, from cells to brains.

The derivation of FTP with the interference term is straightforward [3]. Simply define the interference coefficients as

$$\lambda_i = \frac{P(b=\beta_i|C) - P(a=\alpha_1|C)P(b=\beta_i|C_{\alpha_1}) - P(a=\alpha_2|C)P(b=\beta_i|C_{\alpha_2})}{2\sqrt{P(a=\alpha_1|C)P(b=\beta_i|C_{\alpha_1})P(a=\alpha_2|C)P(b=\beta_i|C_{\alpha_2})}} \tag{4.14}$$

and then solve this equality with respect to the probability $P(b=\beta_i|C)$.

The meaning of the nominator is clear: it is the difference between two types of contextual probabilities, one can say "interference" between these contexts. The denominator gives a proper normalization, cf. Sect. 4.1.1.

In Sect. 4.1.1, the FTP with the interference term was derived by using the quantum wave function, in particular, the phase θ was related to the phase of the wave function. Surprisingly, one can proceed the other way around and reconstruct the corresponding quantum state ψ, the "wave function", from the FTP with the interference term, see [3].

On Double Stochasticity of Contextual Probabilities

As we have seen, the coefficient of interference given by (4.13) need not be bounded by one. This is one of the reasons to use generalizations of quantum formalism based on lifting maps and POVMs. In this section we present another reason for departure from the usage of the standard quantum observables given by Hermitian operators. Consider the quantum formalism in the two dimensional space. There are given two quantum observables a and b represented by Hermitian matrices (2×2), $\widehat{a}$ and $\widehat{b}$ with eigenvalues α_1, α_2 and β_1, β_2, respectively. These are values which can be obtained in measurements of these observables. Let $C_i^b = \{b = \beta_i\}$. Consider contextual probabilities $p_{ji} = P(a = \alpha_j | C_i^b)$. Thus we measure the a-observable under the condition that the b-observable was measured and the result $b = \beta_i$ was obtained. These probabilities are also known as *transition probabilities* (Sect. 4.1.1). We remark that, for any context C, $P(a = \alpha_1|C) + P(a = \alpha_2|C) = 1$. (The total probability to get one of the values of a under the fixed context C equals one.) Hence,

$$p_{1i} + p_{2i} = 1, \quad i = 1, 2. \tag{4.15}$$

This property is called *stochasticity* of transition probabilities. The matrix of transition probabilities (p_{ji}) is called stochastic. Since for any fixed context, contextual probabilities are described by the standard classical probabilistic model (of Kolmogorov), quantum transition probabilities (a special case of contextual probabilities) are also constrained by (4.15). However, they even satisfy the following equalities:

$$P(a = +1|C_1^b)+P(a = +1|C_2^b) = 1, \; P(a = -1|C_1^b)+P(a = -1|C_2^b) = 1 \tag{4.16}$$

or, for matrix elements,

$$p_{j1} + p_{j2} = 1, j = 1, 2. \tag{4.17}$$

The combination of the constraints (4.15) and (4.16) is known as *double stochasticity*. (The matrix of transition probabilities (p_{ji}) is called doubly stochastic.) This is essentially a quantum property. In classical probability theory double stochasticity can be violated. Here typically we have only stochasticity of transition probabilities.

However, biological systems can produce contextual (transition) probabilities, which violate the condition of double stochasticity, see, e.g., Sect. 5.1.2 and also [5–9] it was pointed out that the well known statistical data on the disjunction effect collected by Shafir and Tversky [10–12] violate this condition. Hence, sometimes the conventional quantum observables given by Hermitian matrices cannot be applied. More general quantum observables, see Sect. 4.1.1 for discussion, have to be used.

At the end of this section, we explain the origin of the term *transition probabilities* and derive double stochasticity of quantum transition probabilities.

Let (e_j^a) and (e_i^b) be two orthonormal bases consisting of eigenvectors of the operators $\widehat{a}$ and $\widehat{b}$. Thus $\widehat{a}e_j^a = \alpha_j e_j^a$ and $\widehat{b}e_i^b = \beta_i e_i^b$. By the projection postulate

(Sect. 2.2.2) after measurement of b with the result $b = \beta_i$ the initial state ψ of a quantum system is projected onto the eigenstate e_i^b of $\widehat{b}$ corresponding to the eigenvalue β_i. Now consider this eigenstate as the initial state and perform the a-measurement. Suppose that $a = \alpha_j$. Again by the projection postulate it is projected onto the eigenstate e_j^a of $\widehat{a}$ corresponding to its eigenvalue α_j. This process can be formally considered as transition from the input-state e_i^b to the output-state e_j^a. The probability of such transition is equal to $p_{ji} = P(a = \alpha_j | b = \beta_i)$.

We can express transition probabilities via eigenvectors of $\widehat{a}$ and $\widehat{b}$. Since for the successive a-measurement the output of the previous b-measurement, namely, $\psi_{\text{out}} = e_i^b$, is considered as the input state, we can apply the basic probabilistic rule of quantum mechanics, Born's rule, see the Eq. (2.33), in the following form: $p_{ji} = |\langle \psi_{\text{out}}, e_j^a \rangle|^2 = |\langle e_i^b, e_j^a \rangle|^2$. Hence, $p_{j1} + p_{j2} = |\langle e_1^b, e_j^a \rangle|^2 + |\langle e_2^b, e_j^a \rangle|^2 = \langle e_j^a, e_j^a \rangle = 1$.

The same terminology is used in the theory of *Markov chains*. However, in this theory states are not given by normalized vectors of complex Hilbert space and probabilities are not defined with the aid of Born's rule (thus there is no reason for their double stochasticity).

4.1.2 Conditional Probability and Joint Probability in Quantum Systems

In the previous section, we introduced contextual probabilities for two incompatible observables a and b. This context dependency makes an example of non-Kolmogorovness. It is also known that quantum probability is another example of non-Kolmogorov probability. The conditional probability and the joint probability do not generally exist in quantum systems, which is an essential difference from classical system. In this section, we discuss the existence of conditional probability or joint probability within the mathematical framework of quantum mechanics.

Let $\mathscr{H}$, $\mathscr{K}$ be the Hilbert spaces describing the system of interest, let $\mathscr{S}(\mathscr{H})$ be the set of all states or probability measures on $\mathscr{H}$, let $\mathscr{O}(\mathscr{H})$ be the set of all observables or events on $\mathscr{H}$, and let $\mathscr{P}(\mathscr{H}) \subset \mathscr{O}(\mathscr{H})$ be the set of projections in $\mathscr{O}(\mathscr{H})$.

In classical probability (introduced in Sect. 2.1), the joint probability for two events A and B is expressed as $\mu(A \cap B)$ with a probability measure μ. And the conditional probability is defined by

$$\frac{\mu(A \cap B)}{\mu(B)}.$$

In quantum probability, if the von Neumann-Lüder projection rule is correct, after a measurement of $F \in \mathscr{P}(\mathscr{H})$, a state ρ is considered to be

$$\rho_F = \frac{F \rho F}{\mathrm{tr} \rho F}.$$

When we observe an event $E \in \mathscr{P}(\mathscr{H})$, the expectation value becomes

$$\mathrm{tr}\rho_F E = \frac{\mathrm{tr}F\rho FE}{\mathrm{tr}\rho F} = \frac{\mathrm{tr}\rho FEF}{\mathrm{tr}\rho F}. \tag{4.18}$$

This expectation value can be a candidate of the *conditional probability in quantum probability* (*QP*).

There is another candidate for the conditional probability in QP, which is a direct generalization of classical probability (CP). These alternative expressions of joint probability and the conditional probability in QP are expressed as

$$\varphi(E \wedge F) \quad \text{and} \quad \frac{\varphi(E \wedge F)}{\varphi(F)}, \tag{4.19}$$

where φ is a state (a measure) and $\wedge$ is the meet of two events (projections) corresponding to $\cap$ in CP, and for the state describing by a density operator, we have

$$\varphi(\cdot) = \mathrm{tr}\rho(\cdot).$$

We ask when the above two expressions (4.18) and (4.19) in QP are equivalent. From the next proposition, $\varphi(\cdot \wedge F)/\varphi(F)$ is not a probability measure on $\mathscr{P}(\mathscr{H})$.

Proposition 4.1 (1) *When E commutes with F, i.e. $EF = FE$, the above two expressions are equivalent, namely,*

$$\frac{\varphi(FEF)}{\varphi(F)} = \frac{\varphi(E \wedge F)}{\varphi(F)}.$$

(2) *When $EF \neq FE$, $\frac{\varphi(\cdot \wedge F)}{\varphi(F)}$ is not a probability on $\mathscr{P}_{\mathscr{H}}$, so that the above two expressions are not equivalent.*

Proof (1) $EF = FE$ implies $E \wedge F = EF$ and $FEF = EFF = EF^2 = EF$, so that

$$\frac{\varphi\,(E \wedge F)}{\varphi\,(F)} = \frac{\varphi\,(FEF)}{\varphi\,(F)} = \frac{\varphi\,(EF)}{\varphi\,(F)}.$$

(2) Put $K_\varphi\,(E|F) \equiv \frac{\varphi(E \wedge F)}{\varphi(F)}$ and put $z \in \mathrm{linsp}\,\{x, y\}\,, z \neq x, y$ for any $x, y \in \mathscr{H}$. Take the projections $P_x = |x\rangle\,\langle x|\,,\ P_y = |y\rangle\,\langle y|\,,\ P_z = |z\rangle\,\langle z|$ such that $\left(P_x \vee P_y\right) \wedge P_z = P_z$ and $P_x \wedge P_z = 0 = P_y \wedge P_z$. Then

$$K_\varphi\left(P_x \vee P_y \mid P_z\right) = K_\varphi\left(P_z \mid P_z\right) \neq 0, \quad K_\varphi(P_x \mid P_z) + K_\varphi\left(P_y \mid P_z\right) = 0.$$

Therefore,

$$K_\varphi\left(P_x \vee P_y \mid P_z\right) \neq K_\varphi\left(P_x \mid P_z\right) + K_\varphi\left(P_y \mid P_z\right)$$

so that $K_\varphi\,(\cdot \mid P_z)$ is not a probability measure on $\mathscr{P}(\mathscr{H})$.

In CP, the joint distribution for two random variables f and g is expressed as

$$\mu_{f,g}(\Delta_1, \Delta_2) = \mu\left(f^{-1}(\Delta_1) \cap g^{-1}(\Delta_2)\right)$$

for any Borel sets $\Delta_1, \Delta_2 \in B(\mathbb{R})$. The corresponding quantum expression is either

$$\varphi_{A,B}(\Delta_1, \Delta_2) = \varphi(E_A(\Delta_1) \wedge E_B(\Delta_2)) \quad \text{or} \quad \varphi(E_A(\Delta_1) \cdot E_B(\Delta_2))$$

for two observables A, B and their spectral measures $E_A(\cdot)$, $E_B(\cdot)$ such that

$$A = \int aE_A(da), \; B = \int bE_B(da).$$

It is easily checked that the above expressions do not satisfy either the condition of probability measure or the marginal condition unless $AB = BA$, so that they can not be the joint quantum probability in the classical sense.

Let us explain the above situation, as an example, in a physical measurement process. When an observable A has a discrete decomposition like

$$A = \sum_k a_k F_k, \; F_i \perp F_j \; (i \neq j),$$

the probability obtaining a_k by measurement in a state ρ is

$$p_k = \text{tr}\rho F_k$$

and the state ρ is changed to a (conditional) state ρ_k such that

$$\rho_k = \frac{F_k \rho F_k}{\text{tr}\rho F_k}.$$

After the measurement of A, we will measure a similar type observable B i.e., $B = \sum_j b_j E_j$, $\left(E_i \perp E_j \; (i \neq j)\right)$ and the probability obtaining b_j, after we have obtained the above a_k for the measurement of A, is given by

$$p_{jk} = (\text{tr}\rho F_k)\left(\text{tr}\rho_k E_j\right) = \text{tr}\rho F_k E_j F_k = P_\rho\left(E_j|F_k\right)\text{tr}\rho F_{k.}, \tag{4.20}$$

where $P_\rho(E_j|F_k) = \text{tr}\rho_k E_j$. This p_{jk} satisfies

$$\sum_{j,k} p_{jk} = 1, \; \sum_j p_{jk} = \text{tr}\rho F_k = p_k, \tag{4.21}$$

but not

$$\sum_k p_{jk} = \text{tr}\rho E_j$$

unless $E_j F_k = F_k E_j$ $(\forall j, k)$ so that p_{jk} is not considered as a joint quantum probability distribution. More intuitive expression breaking the usual classical probability law is the following:

$$p_{jk} = P(B = b_j|A = a_k)P(A = a_k) \text{ and } P(B = b_j) \neq \sum_k P(B = b_j|A = a_k)P(A = a_k)$$

Therefore we conclude that in quantum system the above two candidates cannot satisfy the properties of both conditional and joint probabilities in the sense of classical system.

The above discussion shows that the order of the measurements of two observables A and B is essential and it gives us a different expectation value, hence the state changes.

4.1.3 Proposal of Non-Kolmogorov Probability Theory

As we have seen, in the two slit experiment the basic law of classical probability theory, FTP, is violated. Thus we have to apply another probability model, different from the conventional Kolmogorov model.

As was pointed out in the preface, the situation in probability theory can be compared with the situation in geometry. We know well that, besides the Euclidean geometry, there exist various non-Euclidean geometries. The first non-Euclidean model was proposed by Lobachevsky; it is also known as the hyperbolic geometry. We recall the parallel postulate of the Euclidean geometry. By this postulate in two-dimensional Euclidean space, for any line L and point P which does not belong to L , there exists precisely one line going through the point P that does not intersect L.; i.e., that is parallel to R. In hyperbolic geometry there exist at least two distinct lines going through the point P which do not intersect L. Thus the parallel postulate is violated. We also remark that in general we can make the sum of inner angles of triangle as we like. For instance, this sum can be less than 180° or more than 180. Therefore we have many different non-Euclidean geometries. This entails the non-Kolmogorov probability theory versus the Kolmogorov probability theory, that is, we have many different non-Kolmogorov probability theories.

Such non-Kolmogorov probability models will be discussed in the following chapters. The dynamics in non-Kolmogorov probability model can be mathematically described by the adaptive dynamics discussed in the Sect. 4.4.

4.2 Lifting and Channel

In this section, we present the notions of *lifting* [13] and *channel* which are advanced tools of quantum information [14]. They will play a basic role in AD, see Sect. 2.6 for the preliminary discussion. For simplicity, we proceed in the finite dimensional case.

Let $\mathcal{H}$ be a complex Hilbert space. Denote the spaces of linear operators acting in this space by the symbol $\mathcal{O}(\mathcal{H})$. (Thus $A : \mathcal{H} \to \mathcal{H}$, linearity means that A transforms linear combinations of vectors into linear combinations. By fixing an orthonormal basis in $\mathcal{H}$ we can represent it as $\mathbb{C}^n$ and $\mathcal{O}(\mathcal{H})$ as the space of all $n \times n$ complex matrices. However, we prefer to use the operator terminology, which can be applied even to infinite dimensional Hilbert spaces.)[2] The space of all quantum states, i.e., density operators in $\mathcal{H}$, see Sect. 2.2.1 for definition, is denoted by the symbol $\mathcal{S}(\mathcal{H})$.

Let $\mathcal{H}$ and $\mathcal{K}$ be two complex Hilbert spaces and let $\mathcal{H} \otimes \mathcal{K}$ be their tensor product (Sect. 2.5). In many applications the first Hilbert space is the state space of a system under investigation, the second represents states of an environment (bath) and the third is the state space of the compound system. However, for a moment, we do not make such a specification and just consider these spaces as state spaces of two systems (physical, biological, social,...) and the compound system formed of them. We shall also use $\mathcal{O}(\mathcal{H} \otimes \mathcal{K})$ as the space of observables (see the above footnote) for the compound system.

Definition Quantum-like lifting is a map

$$\mathcal{E}^* : \mathcal{S}(\mathcal{H}) \to \mathcal{S}(\mathcal{H} \otimes \mathcal{K}).$$

A Lifting is called pure, if it maps pure states into pure states.

We shall also use a very general definition of a quantum channel:

Definition Quantum-like channel is a map

$$\Lambda^* : \mathcal{S}(\mathcal{H}) \to \mathcal{S}(\mathcal{H}),$$

A channel is called pure, if it maps pure states into pure states.

In the same way we define a channel from $\mathcal{S}(\mathcal{H})$ to $\mathcal{S}(\mathcal{K})$. We also remark that, in fact, each channel can be represented as a lifting by selecting one of Hilbert spaces as the one dimensional complex space $\mathbb{C}$. The reader can check this by taking into account that, for an arbitrary Hilbert space $\mathcal{H}$, the tensor product $\mathcal{H} \otimes \mathbb{C}$ can be identified with $\mathcal{H}$.

Typically one uses only linear maps (since linearity is considered as one of the main distinguishing features of quantum mechanics) and, besides positivity, i.e., mapping positively defined operators into positively defined operators (and density operators are positively defined), one imposes a stronger restriction on the class of channel-maps—the so called complete positivity. Biological QL-models need not be linear, therefore we do not restrict channels (or liftings) to linear maps. The

[2] In literature on quantum mathematical physics, especially in various mathematical models of algebraic quantization, elements of $\mathcal{O}(\mathcal{H})$ are called *observables*, although in physical literature this terminology is reserved for Hermitian operators. In this section we proceed by calling elements of $\mathcal{O}(\mathcal{H})$ quantum observables.

condition of complete positivity is debatable even in standard quantum information theory. There is no reason to use it in biology. Hence, the classes of QL liftings and channels are essentially wider than the conventional ones. We present some important examples of liftings.

Example 4.1 Isometric lifting: Let $V : \mathscr{H} \to \mathscr{H} \otimes \mathscr{K}$ be an isometry, i.e., a linear map satisfying the restriction

$$V^* V = I,$$

where $I \equiv I_{\mathscr{H}}$ is the unit map in $\mathscr{H}$ and V^* denotes the adjoint operator (Sect. 2.3, see (2.49)). Then the map

$$\mathscr{E}^* \rho = V \rho V^*$$

is a lifting. Liftings of this type are called isometric. Every isometric lifting is a pure lifting.

Example 4.2 Reduction (Open system dynamics): If a system Σ_1 interacts with an external system Σ_2 described by another Hilbert space $\mathscr{K}$ and the initial states of Σ_1 and Σ_2 are ρ_1 and ρ_2, respectively, then the combined state θ_t of Σ_1 and Σ_2 at time t after the interaction between the two systems is given by

$$\theta_t \equiv U_t(\rho_1 \otimes \rho_2) U_t^*,$$

where $U_t = \exp(-itH)$ with the total Hamiltonian H of Σ_1 and Σ_2. A channel is obtained by taking the partial trace w.r.t. $\mathscr{K}$ such as

$$\rho_1 \to \Lambda^* \rho_1 \equiv \mathrm{tr}_{\mathscr{K}} \theta_t.$$

Example 4.3 The projection quantum measurement process is written, in the terminology of lifting, as follows. Any observable can be represented as $A = \sum_n a_n \pi_n$, where $\pi_n, n = 1, \ldots, m$, are the projectors on the eigensubspaces corresponding to the eigenvalues a_n (we state again that we work in the finite dimensional case). In accordance with the projection postulate (in the Lüders form, see Sect. 2.2.2, (2.36)) the postmeasurement state (if the result of measurement is not specified) will be

$$\Lambda^* \rho = \sum_n \pi_n \rho \pi_n,$$

where ρ is the input state. This is a channel. Now we fix a discrete probability μ. It is given by a sequence of weights, say $\mu_n \geq 0, \sum_n \mu_n = 1$. The lifting $\mathscr{E}^*$ associated with this channel and with a fixed decomposition of ρ as

$$\rho = \sum_n \mu_n \rho_n$$

where $\rho_n \in \mathscr{S}(\mathscr{H})$ is given by

$$\mathscr{E}^*\rho = \sum_n \mu_n \rho_n \otimes \Lambda^* \rho_n.$$

4.3 Adaptive Dynamics

4.3.1 Motivation and Examples

The methodology of modern science is heavily based on the "Reductionism". To understand a physical system, one has to divide it to simpler subsystems, to study their inter-relations and combinations and in this way to understand the whole system.

Our method about "how to see objects (existence)" is based on the standard methodology of reductionism. The way "how to see objects" is strongly related to setting the mode for observation, such as selection of phenomena and operation for recognition. Our method is called "*Adaptive dynamics (AD)*" *or* "*Adaptive scheme*" for understanding the existence.

In this section, we discuss the conceptual frame of AD from theories of chaos and quantum algorithms which can be considered as the first steps toward creation complete mathematical theory of adaptivity.

4.3.2 Conceptual Meaning

The adaptive dynamics has two aspects, one of which is the "observable-adaptivity" and another is the "state-adaptivity".

The idea of observable-adaptivity comes from the papers [15–17] studying chaos. We claimed that the description of real observations does not match completely with mathematical universalities such as taking limits, sup, inf,etc. Observation of chaos is the result of taking into account suitable scales of, for example, time and space. In the limiting case such an observation is impossible, i.e., chaos will disappear.

The idea of the state-adaptivity is implicitly started in the construction of a compound state for quantum communication [14, 18–20] and in Accardi's Chameleon dynamics [21].[3] This adaptivity can be used to solve a problem pending for more than 30 years: whether there exists an algorithm solving a NP complete problem in polynomial time. We found such algorithms firstly by quantum chaos algorithm [22, 23] and secondly by the adaptive dynamics [13] based on quantum algorithm of the SAT [24, 25].

---

[3] This dynamics provides a possibility to express the essentials of quantum measurements to adaptivity. In particular, Accardi was able to model violation of the famous Bell inequality with the aid of this dynamics.

4.4 State and Observable Adaptive Dynamics

In the Chap. 1 we discussed the intuitive meaning of the adaptive dynamics, and we pointed out that there exist two types of adaptive dynamics; *the observable-adaptive dynamics and the state-adaptive dynamics.*

The observable-adaptive dynamics is a dynamics characterized as follows:

(1) *Measurement depends on how to see an observable to be measured.*
(2) *The interaction between two systems depends on how a fixed observable exists, that is, the interaction is related to some aspects of observables to be measured or prepared.*

The state-adaptive dynamics is a dynamics characterized as follows:

(1) *Measurement depends on how the state and observable to be used exist.*
(2) *The correlation between two systems depends on how the state of at least one of the systems exists, e.g., the interaction Hamiltonian depends on the state.*

In Sect. 1.4, we explained the adaptivity of biological systems, and introduced the concept of the adaptive dynamics. In this section, we will discuss how to use the concept of lifting to explain phenomena breaking the usual probability law.

The adaptive dynamics implies that the dynamics of a state or an observable after an instant (say the time t_0) attached to a system of interest is affected by the existence of some other observable and state at that instant. Let $\rho \in \mathcal{S}(\mathcal{H})$ and $A \in \mathcal{O}(\mathcal{H})$ be a state and an observable before t_0, and let $\sigma \in \mathcal{S}(\mathcal{H} \otimes \mathcal{K})$ and $Q \in \mathcal{O}(\mathcal{H} \otimes \mathcal{K})$ be a state and an observable to give an effect to the state ρ and the observable A. In many cases, the effect to the state is dual to that to the observable, so that we will only discuss the effect to the state. This effect is described by a lifting $\mathcal{E}^*_{\sigma Q}$, so that the state ρ becomes $\mathcal{E}^*_{\sigma Q}\rho$ first, then it will be $\mathrm{tr}_{\mathcal{K}}\mathcal{E}^*_{\sigma Q}\rho \equiv \rho_{\sigma Q}$. The adaptive dynamics here is the whole process such as

$$\textit{Adaptive Dynamics}: \ \rho \Rightarrow \mathcal{E}^*_{\sigma Q}\rho \Rightarrow \rho_{\sigma Q} = \mathrm{tr}_{\mathcal{K}}\mathcal{E}^*_{\sigma Q}\rho$$

That is, what we need is how to construct the lifting for each problem to be studied, that is, we properly construct the lifting $\mathcal{E}^*_{\sigma Q}$ by choosing σ and Q properly. The expectation value of another observable $B \in \mathcal{O}(\mathcal{H})$ in the adaptive state $\rho_{\sigma Q}$ is

$$\mathrm{tr}\rho_{\sigma Q}B = \mathrm{tr}_{\mathcal{H}}\mathrm{tr}_{\mathcal{K}}B \otimes I\mathcal{E}^*_{\sigma Q}\rho.$$

Now suppose that there are two quantum event systems

$$\mathcal{A} = \{a_k \in \mathbb{R}, F_k \in \mathcal{P}(\mathcal{H})\},$$
$$\mathcal{B} = \{b_j \in \mathbb{R}, E_j \in \mathcal{P}(\mathcal{K})\},$$

where we do not assume F_k, E_j are projections, but they satisfy the conditions $\sum_k F_k = I$, $\sum_j E_j = I$ as positive operator valued measure (POVM) corresponding to the partition of a probability space in classical system.[4] Then the "joint-like" probability obtaining a_k and b_j might be given by the *formula*

$$P(a_k, b_j) = \text{tr} E_j \boxdot F_k \mathcal{E}^*_{\sigma Q} \rho, \tag{4.22}$$

where $\boxdot$ is a certain operation (relation) between A and B, more generally, one can take a certain operator function $f(E_j, F_k)$ instead of $E_j \boxdot F_k$. If σ, Q are independent from any F_k, E_j and the operation $\boxdot$ is the usual tensor product $\otimes$ so that A and B can be considered in two independent systems or to be commutative, then the above "joint-like" probability becomes the joint probability. However, if this is not the case, e.g., Q is related to A and B, the situation will be more subtle. Therefore, the problem is how to set the operation $\boxdot$ and how to construct the lifting $\mathcal{E}^*_{\sigma Q}$ in order to describe the particular problems associated to systems of interest. In the sequel, we will discuss this problem in the context dependent systems like bio-systems and psyco-systems mentioned in Introduction. That is, we discuss how to apply formula (4.22) to the following three problems breaking the usual probability law: (1) State change of tongue for sweetness, (2) Lactose-glucose interference in *E. coli* growth, (3) Updating the Bayesian law.

References

1. Feynman, R., Hibbs, A.: Quantum Mechanics and Path Integrals. McGraw-Hill, New York (1965)
2. Khrennikov, A.: Interpretations of Probability, 2nd edn. De Gruyter, Berlin (2009)
3. Khrennikov, A.: Contextual Approach to Quantum Formalism. Springer, Berlin (2009)
4. Asano, M., Basieva, I., Khrennikov, A., Ohya, M., Yamato, I.: Non-Kolmogorovian approach to the context-dependent systems breaking the classical probability law. Found. Phys. **43**(7), 895–911 (2013)
5. Conte, E., Todarello, O., Federici, A., Vitiello, F., Lopane, M. and Khrennikov, A.: A preliminary evidence of quantum-like behaviour in measurements of mental states. In: Quantum Theory: Reconsideration of Foundations. Series Mathematical Modelling in Physics, Engineering and Cognitive Sciences, vol. 10, pp. 679–702. Växjö University Press, Växjö (2004)
6. Conte, E., Todarello, O., Federici, A., Vitiello, F., Lopane, M., Khrennikov, A., Zbilut, J.P.: Some remarks on an experiment suggesting quantum-like behavior of cognitive entities and formulation of an abstract quantum mechanical formalism to describe cognitive entity and its dynamics. Chaos, Solitons Fractals **31**(5), 1076–1088 (2007)
7. Conte, E., Khrennikov, A., Todarello, O., Federici, A., Zbilut, J.P.: A preliminary experimental verification on the possibility of Bell inequality violation in mental states. Neuroquantology **6**, 214–221 (2008)
8. Conte, E., Khrennikov, A., Todarello, O., Federici, A., Zbilut, J.P.: Mental states follow quantum mechanics during perception and cognition of ambiguous figures. Open Syst. Inform. Dyn. **16**, 1–17 (2009)
9. Khrennikov, A.: Ubiquitous Quantum Structure: From Psychology to Finance. Springer, Heidelberg (2010)

[4] See Chap. 10 for the detailed presentation of theory of POVMs and their applications in cognitive science and decision making.

10. Tversky, A., Shafir, E.: The disjunction effect in choice under uncertainty. Psychol. Sci. **3**, 305–309 (1992)
11. Shafir, E., Tversky, A.: Thinking through uncertainty: nonconsequential reasoning and choice. Cogn. Psychol. **24**, 449–474 (1992)
12. Croson, R.: The disjunction effect and reasoning-based choice in games. Org. Behav. Hum. Decis. Process. **80**, 118–133 (1999)
13. Accardi, L., Ohya, M.: A Stochastic limit approach to the SAT problem. Open Syst. Inf. Dyn. **11**, 1–16 (2004)
14. Accardi, L., Ohya, M.: Compound channels, transition expectations, and liftings. Appl. Math. Optim. **39**, 33–59 (1999)
15. Kossakowski, A., Ohya, M., Togawa, Y.: How can we observe and describe chaos? Open Syst. Inf. Dyn. **10**(3), 221–233 (2003)
16. Ohya, M.: Foundation of chaos through observation. In: Hida, T., Saito, K., Si, S. (eds.) Quantum Information and Complexity, pp. 391–410 (2004)
17. Ohya, M.: Adaptive dynamics in quantum information and chaos. In: Hida, T. (ed.) Stochastic Analysis: Classical and Quantum, pp. 127–142 (2005)
18. Ohya, M.: On compound state and mutual information in quantum information theory. IEEE Trans. Inf. Theory **29**, 770–777 (1983)
19. Ohya, M.: Some aspects of quantum information theory and their applications to irreversible processes. Rep. Math. Phys. **27**, 19–47 (1989)
20. Ohya, M.: Note on quantum probability. Il Nuovo Cimento **38**, 203–206 (1983)
21. Accardi, L.: Urne e camaleonti: Dialogo sulla realta, le leggi del caso e la teoria quantistica.Il Saggiatore, Rome (1997); English edn., World Scientific, Singapore (2002); Japanese edition, Makino, Tokyo (2002); Russian edn., Regular and Chaotic dynamics, Ijevsk (2002)
22. Ohya, M., Volovich, I.V.: New quantum algorithm for studying NP-complete problems. Rep. Math. Phys. **52**, 25–33 (2003)
23. Ohya, M., Volovich, I.V.: Mathematical Foundations of Quantum Information and Computation and Its Applications to Nano- and Bio-systems. Springer, Berlin (2011)
24. Ohya, M., Masuda, N.: NP problem in quantum algorithm. Open Syst. Inf. Dyn. **7**(1), 33–39 (2000)
25. Accardi, L., Sabbadini, R.: On the Ohya-Masuda quantum SAT Algorithm. In: Proceedings International Conference UMC'01. Springer (2001)

Chapter 5
Application of Adaptive Dynamics to Biology

Abstract In this chapter, we present examples of the nonclassical probabilistic behavior of concrete microbiological systems: cells and proteins. The experimental data exhibit the interference effect which is similar to interference of probabilities observed in quantum physics. Finally, we apply the formalism of quantum adaptive dynamics to model nonclassical probabilistic behavior of biosystems.

Keywords Adaptive dynamics · Molecular biology · Quantum-like models · Lactose-glucose interference in *Escherichia coli* growth · Differentiation of a tooth stem cell · Mesenchymal cells · PrP^C and PrP^{Sc} prions

In this chapter, we illustrate by a few examples of cell and protein behavior that microscopic bio-systems can exhibit complex probabilistic behavior, which cannot be described by classical probabilistic laws. These examples support authors' conjecture that the behavior of microscopic bio-systems have to be described by QL-models, i.e., models based on the formalism of quantum-mechanics, see Chap. 1. As is emphasized in this chapter, we do not couple QL-behavior with quantum physical processes in bio-systems. We present arguments that such a behavior can be induced by the information complexity of even smallest bio-systems, by their adaptivity to context changes.[1] Although our examples of the QL-behavior are rather simple, lactose-glucose interference in *E. coli* growth, the interference effect for the differentiation of a tooth *stem cell* induced by the presence of *mesenchymal cells*, interference in behavior of PrP^C and PrP^{Sc} *prions*, these examples can stimulate the interest to QL-models of adaptive dynamicsand lead to more complex examples of nonclassical probabilistic behavior in cellular biology.

We concentrate our modeling mainly on the glucose effect on *E. coli* growth. This is a basic example of destructive interference in cell's activity, the interference of

[1] It is well known that the behavior of each cell is characterized by huge complexity. It can be compared with an actor who is playing simultaneously at a few different scenes participating in different (sometimes mutually incompatible) performances. Even smaller bio-systems, such as viruses and even proteins, exhibit complex behavior, which is characterized by nontrivial context-dependence and adaptivity to context variations.

M. Asano et al., *Quantum Adaptivity in Biology: From Genetics to Cognition*,
DOI 10.1007/978-94-017-9819-8_5

two factors: the presence of lactose and glucose in a *E. coli* cell.[2] One of the reasons for this is the possibility to find quantitative statistical data for varying experimental contexts, the presence of only lactose, only glucose, and both lactose and glucose,[3] see, e.g., Inada et al. [2].

5.1 Violation of the Formula of Total Probability in Molecular Biology

We shall use the contextual viewpoint to conditioning and FTP (the formula of total probability) presented in Sect. 4.1.1.

5.1.1 Reaction of Tongue to Sweetness

The first problem under investigation is not so sophisticated, but quite common. It has a heuristically simple contextual structure. We consider the following cognitive experiment.

One takes sugar S or (and) chocolate C and he is asked whether it is sweet or not. The answers "yes" and "no" are numerically encoded by 1 and 2. Then the basic classical probability law need not be satisfied, that is,

$$P(C=1) \neq P(C=1|S=1)P(S=1) + P(C=1|S=2)P(S=2),$$

because the LHS $P(C=1)$ will be very close to 1, but the RHS will be less than $\frac{1}{2}$. Note that the LHS $P(C=1)$ is obtained in the context that subjects do not taste sugar; they start directly with chocolate. The contexts "a tongue tasted sugar", say S_{sug}, and "a tongue did not taste sugar", say $S_{\neg sug}$, are different. The probabilities in LHS and RHS in the above equation should be replaced by $P_{S_{\neg sug}}(C=1)$ and $P_{S_{sug}}(C=1)$. The problem to be discussed is how to obtain these probabilities mathematically.

---

[2] Since the operon theory was proposed in 1956–1961 by Jacob and Monod [1], the regulatory system of gene expression of lactose operon has been extensively studied and the molecular mechanism of it has been mostly elucidated including the catabolite repression phenomenon. The description of the molecular biological mechanism so far obtained and also the systems biology approach (reductionism) are presented in Sects. 3.3.1 and 3.3.2.

[3] It is clear that this contextual structure is isomorphic to the contextual structure of the two slit experiment in quantum physics, Sect. 4.1.1.

5.1.2 Glucose Effect on E. coli *Growth*

Our considerations are based on an article reporting the glucose effect on *E. coli* (*Escherichia coli*) growth [2], where the β-galactosidase activity at certain growth phase was measured under the following experimental contexts:

C : 0.4 % lactose + 0.1 % glucose,
C_L : 0.4 % lactose (and no glucose),
C_G : 0.4 % glucose (and no lactose).

The activity is represented in *Miller units* (enzyme activity measurement condition). Then one can obtain the probabilistic data: 0.4 % lactose +0.1 % glucose, 43 units; 0.4 % lactose, 2920 units; 0.4 % glucose, 33 units.

We recall that by full induction, the activity reaches to 3000 units. We want to represent these data in the form of contextual probabilities and put them into the formula of total probability. We introduce the observable $b = \pm 1$, which describes the β-galactosidase activity. We also introduce the observable $a = L, G$ describing the detection of molecules of lactose and glucose. The experimental data provide the contextual (conditional) probabilities:

$$P(b = +1|C) = \frac{43}{3000} \approx 0.014,$$

$$P(b = +1|C_L) = \frac{2920}{3000} \approx 0.973, \ P(b = +1|C_G) = \frac{33}{3000} \approx 0.011.$$

We also determine the probabilities of "realizations of contexts C_L and C_G in the context C":

$$P(a = L|C) = \frac{0.4}{0.5} = 0.8, \ P(a = G|C) = \frac{0.1}{0.5} = 0.2.$$

In the classical probabilistic framework we would obtain the equality (FTP):

$$P(b = +1|C) = P(a = L|C)P(b = +1|C_L) + P(a = G|C)P(b = +1|C_G). \quad (5.1)$$

By putting the data into (5.1) we obtain

$$0.014 \approx 0.158 \quad (5.2)$$

Thus classical FTP is violated. A cell does not split a context C into two disjoint and not interfering contexts C_L and C_G. It treats C (which is physically just the union of two ensembles of molecules, lactose and glucose)[4] as a new environment.

[4] We ignore the presence of other types of molecules, which are not essential for this type of gene expressions.

We now calculate the coefficient of interference:

$$\lambda_+ = \frac{P(b=+1|C) - P(a=L|C)P(b=+1|C_L) - P(a=G|C)P(b=+1|C_G)}{2\sqrt{P(a=L|C)P(b=+1|C_L)P(a=G|C)P(b=+1|C_G)}},$$

i.e.,

$$\lambda_+ = -1.74. \tag{5.3}$$

This coefficient is larger than one. Since in any way it is nonzero, we cannot proceed with Kolmogorov's model. Since it is larger than one, we cannot proceed with the standard quantum model (see discussion in Sect. 4.1.1). We shall come back to the problem of interference in *E. coli* metabolism in Sect. 5.2.2, where we shall use the generalized quantum calculus generated by QL adaptive dynamics.

Another source of impossibility to apply the standard quantum formalism based on the representation of observables by Hermitian operators is violation of the condition of double stochasticity for contextual probabilities. Contrary to (4.16), $P(b=+1|C_L) + P(a=G|C)P(b=+1|C_G) \neq 1$.

Remark We recall that lactose induces the enzyme, but without induction certain percentage would be expressed by fluctuation of gene expression. The concentration of glucose is not important, but the following should be taken into account:

If we add 0.2 % glucose in the medium, cells can grow to its stationary phase only on glucose and they do not try to utilize lactose. Thus, if we want to see the enzyme induction during the growth, we have to limit the glucose concentration, usually 0.02 %. That amount is insufficient for the support of cell growth and cells try to utilize lactose after the consumption of glucose. If we add only 0.02 % glucose in the medium without any other carbon (energy) source, then the enzyme level would be similar as in the presence of 0.2 % glucose and cells stop growing. If there is any other carbon source than 0.02 % glucose, then cells continue to grow and the enzyme level changes depending on the kind of carbon source (for lactose, the level is quite high; for maltose, the level would be low, but significant; for pepton (amino acid mixture), the level would be a little bit more).

The mathematically oriented reader may be disappointed that the concentration of glucose under the context C, namely, 0.1 %, does not match with its concentration under context the C_G, namely, 0.4 %. However, by taking into account the statement in the second sentence of the above remark we can assume that the level of activation of enzyme would be approximately the same for the experimental context

$$C' : 0.4\,\%\ \text{lactose} + 0.4\,\%\ \text{glucose}.$$

For this context, we can also check FTP under the assumption

$$P(b=+1|C') \approx P(b=+1|C) = \frac{43}{3000} \approx 0.014.$$

Here,

$$P(a = L|C') = P(a = G|C') = 0.5.$$

Hence,

$$\lambda_+ = -2.282.$$

It is again larger than one and we have to use a generalization of the standard quantum formalism.

5.1.3 Mesenchymal Cells Context

It is well known that a dental epithelial cell grows in a medium as it is (no differentiation). A dental mesenchymal cell grows similarly. However, if they are grown together (mix), then a dental epithelial cell differentiates to a tooth cell [3]. The growth condition is the same. Hence, growing independently, they grow as they are (no differentiation). If they grow mixed, dental epithelial cells differentiate.

We introduce a random variable $b = +1$, in the case of differentiation of a cell, and otherwise $b = -1$. We also introduce three contexts:

C : the presence of mesenchymal cells and tooth stem cells;
C_M : the presence of only mesenchymal cells;
C_{ST} : the presence of only tooth stem cells.

Although we do not have the complete statistical data (and this is a general problem in cellular biology), we can assume that $P(b = +1|C_M) = \varepsilon_M \ll 1$ (the probability of differentiation in the ensemble consisting of only mesenchymal cells) and $P(b = +1|C_{ST}) = \varepsilon_{ST} \ll 1$ (the probability of differentiation in the ensemble consisting of only tooth stem cells) and also that $P(b = +1|C) \approx 1$ (the probability of differentiation in a mixture of mesenchymal cells and tooth stem cells). We now introduce another variable a describing the detection of mesenchymal cell, $a = M$, and tooth stem cell, $a = ST$. Now we can take, e.g., a sample of cells with equal concentrations of mesenchymal cells and tooth stem cells, i.e.,

$$P(a = M|C) = P(a = ST|C) = 0.5.$$

Then the classical FTP is violated and the interference coefficient λ_+ is very large.

We can also mention an experiment on the differentiation of $PL1^+$ cells [4]. This experiment was performed in the study of new methods of identification of cancer cells. A population of $PL1^+$ cells was identified by exposing it under two complementary contexts: (a) differentiation context (serum rich media) and (b) stem cells culture context. We can use this experiment as an additional example of destructive interference.

5.1.4 PrP^{Sc} Prion Proteins Context

In spite of its small size, a cell is still an extremely complex biological system. Now we are looking for interference effects and violation of laws of classical probability theory for essentially simpler systems, namely, proteins (which are, although very small and less complex as compared with cells, are still very large and complex at the molecular scale).

Let us consider prion protein (cause of CJD, mad cow disease , see, e.g., [5]). Usually it stays as globular soluble form in mono-dispersed state; this is safe conformation. On heating, it may denature and aggregate taking random conformations. If there exit seeds for β-form conformation, a mutant prion protein, or wild type prion protein (but after prolonged incubation), they tend to make filamentous aggregate (this is the cause of CJD) as a whole, which is resistant to solubilization with detergent and to denaturation by heat. Thus, wild type prion from a normal animal can be incubated for a long time as a clear solution. But by adding mutant prion or rather β-form wild type prion obtained by incubating the wild type prion solution for a very long time, most of the prion proteins become insoluble as filamentous aggregates. After this, the prion proteins can no more resume the initial soluble conformation.

We introduce a random variable $b = +1$, if protein can be incubated for a long time as a clear solution, and otherwise $b = -1$. We also introduce three contexts:

C : the presence of both types of proteins;
C_M : the presence of only mutant prion proteins PrP^{Sc};
C_{PP} : the presence of only wild type prion proteins PrP^C.

Although we again have no sufficiently statistical experimental data, we can assume that $P(b = +1|C_{PP}) \approx 1 - \delta_{PP}$ and $P(b = +1|C_M) \approx \delta_M$ and also that $P(b = +1|C) = \delta$, where all δs are very small.

We also introduce the observable a of detection of protein type, $a = M, PP$. Now we can take, e.g., a sample of proteins of both types with equal concentrations, i.e.,

$$P(a = M|C) = P(a = PP|C) = 0.5.$$

Then FTP is violated with very strong destructive interference.

The same effect, the destructive interference, can be approached by using, instead of mutant proteins, wild type prion protein, but after prolonged incubation.

5.2 Interpretation of the Above Violations by Adaptive Dynamics

We presented some examples of biological phenomena which violate the usual probability law. These violations are of the contextual nature. In this section, we discuss these context dependences from the point of view of adaptive dynamics.

5.2.1 State Change as Reaction of Tongue to Sweetness

In Sect. 5.1.1, we presented one of the simplest examples of adaptive phenomena.

Let $|e_1\rangle$ and $|e_2\rangle$ be the orthogonal vectors describing sweet and non-sweet states, respectively. This problem can be described by the Hilbert space $\mathbb{C}^2$, so that $|e_1\rangle$ and $|e_2\rangle$ can be set as $\binom{1}{0}$ and $\binom{0}{1}$, respectively. *The initial state of a tongue is neutral* such as

$$\rho \equiv |x_0\rangle \langle x_0| ,$$

where $x_0 = \frac{1}{\sqrt{2}} (|e_1\rangle + |e_2\rangle)$. Here the neutral pure state ρ describes the state of tongue before the experiment, and we start from this state.

When one tastes sugar, the state of tongue will be changed. Such change can be written as the operator (in this case, two-dimensional matrix) on the Hilbert space. The operator corresponding to tasting sugar is mathematically (and operationally) represented as

$$S = \begin{pmatrix} \lambda_1 & 0 \\ 0 & \lambda_2 \end{pmatrix},$$

where λ_i are complex numbers satisfying $|\lambda_1|^2 + |\lambda_2|^2 = 1$. This operator can be regarded as the square root of the sugar state σ_S:

$$\sigma_S = |\lambda_1|^2 E_1 + |\lambda_2|^2 E_2, \; E_1 = |e_1\rangle \langle e_1| , \; E_2 = |e_2\rangle \langle e_2| .$$

Taking sugar, he will taste that it is sweet with the probability $|\lambda_1|^2$ and non-sweet with the probability $|\lambda_2|^2$, so $|\lambda_1|^2$ should be much higher than $|\lambda_2|^2$ for usual sugar. This comes from the following change of the neutral initial (i.e., non-adaptive) state of a tongue:

$$\rho \to \rho_S = \Phi_S(\rho) \equiv \frac{S^* \rho S}{\operatorname{tr} |S|^2 \rho}. \tag{5.4}$$

This is the state of a tongue after tasting sugar.

This dynamics is similar to the usual expression of the state change in quantum dynamics. The subtle point of the present problem is that just after tasting sugar the state of a tongue is neither ρ_S nor ρ. Here we note that if we ignore subjectivity (personal features) of one's tongue, then, instead of the state given by (5.4), the "just after tasting sugar" state will have the form

$$E_1 \rho_S E_1 + E_2 \rho_S E_2.$$

This is the unread state (as in quantum measurement, see Sect. 2.2.2). We can use the above two expressions, which give us the same result for computation of the probabilities for the S-variable.

However, for a while the tongue becomes dull to sweetness (and this is the crucial point of our approach for this example), so the tongue state can be written by means of a certain "exchanging" operator $X = \begin{pmatrix} 0 & 1 \\ 1 & 0 \end{pmatrix}$ such that

$$\rho_{SX} = X \rho_S X.$$

Similarly, when one tastes chocolate, the state will be given by

$$\rho_{SXC} = \Phi_C(\rho_{SX}) \equiv \frac{C^* \rho_{SX} C}{\mathrm{tr}\, |C|^2 \rho_{SX}},$$

where the operator C has the form

$$C = \begin{pmatrix} \mu_1 & 0 \\ 0 & \mu_2 \end{pmatrix}$$

with $|\mu_1|^2 + |\mu_2|^2 = 1$. Common experience tells us that $|\lambda_1|^2 \geq |\mu_1|^2 \geq |\mu_2|^2 \geq |\lambda_2|^2$ and the first two quantities are much larger than the last two quantities.

As can be seen from the preceding consideration, in this example the "adaptive set" $\{\sigma, Q\}$ is the set $\{S, X, C\}$. Now we introduce the following (nonlinear) lifting:

$$\mathscr{L}_{\sigma Q}(\rho) = \rho_S \otimes \rho_{SXC} = \Phi_S(\rho) \otimes \Phi_C(X \Phi_S(\rho) X).$$

The corresponding joint probabilities are given by

$$P(S = j, C = k) = \mathrm{tr}\, E_j \otimes E_k \mathscr{L}_{\sigma Q}(\rho).$$

The probability that one tastes sweetness of the chocolate immediately after tasting sugar is

$$P(S = 1, C = 1) + P(S = 2, C = 1) = \frac{|\lambda_2|^2 |\mu_1|^2}{|\lambda_2|^2 |\mu_1|^2 + |\lambda_1|^2 |\mu_2|^2},$$

which is $P_{S_{sug}}(C = 1)$. Note that this probability is much less than

$$P_{\neg S_{sug}}(C = 1) = \mathrm{tr}\, E_1 \Phi_C(\rho) = |\mu_1|^2,$$

which is the probability of sweetness tasted by the neutral tongue ρ. This means that the usual formula of total probability should be replaced by the adaptive (context dependent) probability law.

5.2.2 Activity of Lactose Operon in E. coli

Here we plan to construct AD generating contextual probabilities resulting from the activity of lactose operon and violating FTP, see Sect. 5.1.2. As in this section, we shall operate with contexts C_L, C_G, C, random variables $a = L, G, b = \pm$ and events L and G; L: "*E. coli* detects a lactose molecule in the cell environment to use it for its metabolism" and G: "*E. coli* detects a glucose molecule". In the context C_L the probability $P(L) = 1$ and in the context C_G the probability $P(G) = 1$; in the context C, the probabilities $P(L)$ and $P(G)$ were calculated in Sect. 5.1.2 as $P(L) = 0.8,\ P(G) = 0.2$. Since the AD-formalism generates joint probability, conditional probability can be defined by the Bayes formula. Therefore it is useful to move from the contextual conditioning, which was used in Sect. 5.1.2 (and was very convenient to demonstrate the contextual structure of violation of FTP) to event conditioning. As before, we also consider the events labeled by the symbols $\{+, -\}$: "*E. coli* activates its lactose operon or not". From the experimental data for the contexts C_L and C_G, the following conditional probabilities were obtained (Sect. 5.1.2):

$$\begin{aligned} C_L &: P(+|L) = \frac{2920}{3000}, \\ C_G &: P(+|G) = \frac{33}{3000}. \end{aligned} \tag{5.5}$$

We set $P(+) \equiv P(+|C)$; thus, see Sect. 5.1.2, from the experimental data we obtain

$$P(+) = \frac{43}{3000},$$

We now use our mathematical model for computation of the above probabilities, by using the concept of lifting. First, we introduce the initial state $\rho = |x_0\rangle\langle x_0|$ in the Hilbert space $\mathscr{H} = \mathbb{C}^2$. The state vector x_0 is written as

$$|x_0\rangle = \frac{1}{\sqrt{2}}|e_1\rangle + \frac{1}{\sqrt{2}}|e_2\rangle .$$

The basis $\{e_1, e_2\}$ denote the detection of lactose or glucose by *E. coli*, i.e., the events L and G. In the initial state ρ, the *E. coli* has not recognized the existence of lactose and glucose yet. When *E. coli* recognizes them, the following state change occurs:

$$\rho \mapsto \rho_D = \Phi_D(\rho) \equiv \frac{D\rho D^*}{\mathrm{tr}\,(|D|^2\rho)},$$

where

$$D = \begin{pmatrix} \alpha & 0 \\ 0 & \beta \end{pmatrix}$$

with $|\alpha|^2 + |\beta|^2 = 1$. Note that $|\alpha|^2$ and $|\beta|^2$ give the probabilities of the events L and G: $P(L)$ and $P(G)$. The state $\sigma_D \equiv DD^*$ encodes the probability distribution $P(L), P(G)$:

$$\sigma_D = P(L)\,|e_1\rangle\,\langle e_1| + P(G)\,|e_2\rangle\,\langle e_2|\,.$$

In this sense, the state σ_D represents the chemical solution of lactose and glucose. We call D and ρ_D the *detection operator* and *detection state*, respectively. The state determining the activation of the operon in *E. coli* depends on the detection state ρ_D. In our operational model, this state is obtained as the result of the following transformation:

$$\rho_{DQ} = \Phi_Q(\rho_D) \equiv \frac{Q\rho_D Q^*}{\text{tr}\,(Q\rho Q^*)},$$

where the operator Q is chosen as

$$Q = \begin{pmatrix} a & b \\ c & d \end{pmatrix}.$$

We call ρ_{DQ} and Q the *activation state* for the operon and *activation operator*, respectively. (The components a, b, c, and d will be discussed later.)

We introduce lifting

$$\mathcal{L}_{DQ}(\rho) = \Phi_D(\rho) \otimes \Phi_Q(\Phi_D(\rho)) \in \mathcal{H} \otimes \mathcal{K} = \mathbb{C}^2 \otimes \mathbb{C}^2,$$

by which we can describe the correlation between the activity of lactose operon and the ratio of concentration of lactose and glucose. From the discussion in Sect. 4.4, the joint probabilities $P_{DQ}(L,+)$ and $P_{DQ}(G,+)$ are given by

$$\begin{aligned} P_{DQ}(L,+) &= \text{tr}\,(E_1 \otimes E_1 \mathcal{L}_{DQ}(\rho)), \\ P_{DQ}(G,+) &= \text{tr}\,(E_1 \otimes E_1 \mathcal{L}_{DQ}(\rho)). \end{aligned} \tag{5.6}$$

The probability $P_{DQ}(\pm)$ is obtained as $P_{DQ}(L,\pm) + P_{DQ}(G,\pm)$.

Let us consider a context C_L such that the detection operator D satisfies the condition $P(L) = |\alpha|^2 = 1$. We denote such D by the symbol D_L. The probabilities $P_{D_L Q}(\pm) = P_{C_L}(\pm) (\equiv P(\pm|C_L))$ are calculated as

$$P_{C_L}(+) = \frac{|a|^2}{|a|^2 + |c|^2},\quad P_{C_L}(-) = \frac{|c|^2}{|a|^2 + |c|^2}.$$

From the experimental results, these values should be $\frac{2920}{3000}$ and $\frac{80}{3000}$. Therefore, we can give the following forms for the parameters a and c.

$$a = \sqrt{\frac{2920}{3000}}\mathrm{e}^{\mathrm{i}\theta_{+L}} k_L,\quad c = \sqrt{\frac{80}{3000}}\mathrm{e}^{\mathrm{i}\theta_{-L}} k_L$$

Here, k_L is a certain real number. In a similar way, we consider the context C_G and obtain

$$b = \sqrt{\frac{33}{3000}} \mathrm{e}^{\mathrm{i}\theta_{+G}} k_G, \ d = \sqrt{\frac{2967}{3000}} \mathrm{e}^{\mathrm{i}\theta_{-G}} k_G$$

for the components b and d. To simplify the discussion, hereafter we assume $\theta_{+L} = \theta_{-L}$, $\theta_{+G} = \theta_{-G}$ and denote $\mathrm{e}^{\mathrm{i}\theta_L} k_L$, $\mathrm{e}^{\mathrm{i}\theta_G} k_G$ by $\tilde{k}_L$, $\tilde{k}_G$. Then the operator Q is rewritten as

$$\begin{aligned} Q &= \frac{1}{\sqrt{3000}} \begin{pmatrix} \sqrt{2920} & \sqrt{33} \\ \sqrt{80} & \sqrt{2967} \end{pmatrix} \begin{pmatrix} \tilde{k}_L & 0 \\ 0 & \tilde{k}_G \end{pmatrix} \\ &= \begin{pmatrix} \sqrt{P_{C_L}(+)} & \sqrt{P_{C_G}(+)} \\ \sqrt{P_{C_L}(-)} & \sqrt{P_{C_G}(-)} \end{pmatrix} \begin{pmatrix} \tilde{k}_L & 0 \\ 0 & \tilde{k}_G \end{pmatrix}. \end{aligned} \tag{5.7}$$

By using this Q, we calculate the probability $P_C(+) \equiv P(+|C)$ corresponding to the context C:

$$P_C(+) = \frac{|\sqrt{P_{C_L}(+)}\tilde{k}_L\alpha + \sqrt{P_{C_G}(+)}\tilde{k}_G\beta|^2}{|\sqrt{P_{C_L}(+)}\tilde{k}_L\alpha + \sqrt{P_{C_G}(+))}\tilde{k}_G\beta|^2 + |\sqrt{P_{C_L}(-)}\tilde{k}_L\alpha + \sqrt{P_{C_G}(-)}\tilde{k}_G\beta|^2}. \tag{5.8}$$

In general, the value of this probability is different from that of $P_{C_L}(+)|\alpha|^2 + P_{C_G}(+)|\beta|^2$. The rate $|\tilde{k}_L|/|\tilde{k}_G|$ essentially determines the degree of the difference. Recall the experimental data context C. In this case, $P(L) = |\alpha|^2 = 0.8 > P(G) = |\beta|^2 = 0.2$, but $P_C(+)$ is very small. According to our interpretation, this implies that the rate $|\tilde{k}_L|/|\tilde{k}_G|$ is very small. In this sense, the operator $F = \begin{pmatrix} \tilde{k}_L & 0 \\ 0 & \tilde{k}_G \end{pmatrix}$ in Eq. (5.7) specifies the preference in *E. coli*'s metabolism. We call F the *preference operator*. Finally note that if α, β are real and $\tilde{k}_L = \tilde{k}_G^*$, the formula of total probability holds.

5.3 Precultivation Effect for *E. coli* Growth

In the previous section, we introduced the *E. coli*'s preference $\{k_L/k_G, \theta\}$ based on the adaptive dynamics. In this section, we estimate the preference for a different type of strains. The calculation of preferences is similar to the calculation in the previous section.

Let us explain our experiment to obtain the probabilities $P(+)$. By the method of β-galactosidase assay [6], we measured the values of Miller unit (MU) in five different situations, see Table 5.1. *E. coli* is grown in the media containing (1) 0.4 % lactose; (2) 0.4 % glucose;(3) 0.2 mM IPTG (Isopropyl β-D-1-thiogalactopyranoside); (4) 0.1 % glucose and 0.4 % lactose; (5) 0.4 % glucoseand 0.4 % lactose. In this experiment,

Table 5.1 Results of beta-galactosidase assay

Data	A	B	C	D	E
strain	W3110	W3110	ML30	ML308	ML308-2
lacI	+	+	+	−	−
lacY	+	+	+	+	−
preculturing	LB	Gly	LB	LB	LB
MU(1)	2920†	957	1763	2563	6140
MU(2)	33†	5	14	1592	3062
MU(3)	3000†, 2200‡	1059	3133	2438	5326
MU(4)	43†	486	184	2074	2668
MU(5)	64‡	421	78	2050	862

The symbol '+' or '−' means the genotype of each strain. The values with † are cited from the paper (Inada et al. [2]) and the values with ‡ are cited from the online page http://ro119.com/archive/nagoya.cool.ne.jp/planta/bio/lac_operon.htm

we used four types of *E. coli* [7]; W3110, ML30, ML308 (*lacI*), ML308-2 (*lacI*, *lacY*). We also used two types of preculture condition; grown in minimal medium with 0.4 % glycerol(Gly) or in Luria broth(LB).

The strains of W3110 and ML30 are wild type *E. coli*, which show stronger preference for glucose than for lactose. In fact, as shown in data A and C, both MU(4) and MU(5) have small values similar to MU(2). On the other hand, as seen in data D, the strain ML308 has weak preference for glucose since MU(2), MU(4), and MU(5) are large. ML308 is a mutant of repressor minus ($lacI^-$), so that the β-galactosidase is produced in the cell even in the presence of high concentration of glucose. As seen in data E, ML308-2 shows different behavior from other strains; MU(4) and MU(5) are smaller than MU(2). ML308-2 is a mutant, which is not only repressor minus but also lactose transport system minus ($lacY^-$). Table 5.1 also shows the result for two different preculture conditions as seen in data A and B. We clearly see that the value of MU(4) and MU(5) in data B is very high in comparison to those in data A. This result shows that *E. coli*'s preferences are affected by the preculture condition.

Also we can calculate the conditional probabilities $P(\pm|L)$, $P(\pm|G)$, $P_{(4)}(+)$ and $P_{(5)}(+)$ from the obtained MU values (3)–(5). For data C, we can calculate them as follows.

$$P(+|L) = \frac{1763}{3133}, P(+|G) = \frac{14}{3133},$$
$$P_{(4)}(+) = \frac{184}{3133}, P_{(5)}(+) = \frac{78}{3133}.$$

The Table 5.2 shows the probabilities for data (A)–(E).

One can confirm that the probabilities in data A and C do not satisfy FTP;

$$P(+) \neq P(+|L)P(L) + P(+|G)P(G).$$

Table 5.2 Calculated probabilities for data A, B, C, D and E

	A	B	C	D	E
$P_{(4)}(L)$	0.8	0.8	0.8	0.8	0.8
$P_{(5)}(L)$	0.5	0.5	0.5	0.5	0.5
$P(+\|L)$	0.973	0.904	0.563	1.000	1.000
$P(+\|G)$	0.011	0.005	0.005	0.653	0.575
$P_{(4)}(+)$	0.014	0.459	0.059	0.851	0.501
$P_{(5)}(+)$	0.029	0.398	0.025	0.841	0.462
$P(+\|L)P_{(4)}(L)+$ $P(+\|G)P_{(4)}(G)$	0.780	0.723	0.117	0.722	0.660
$P(+\|L)P_{(5)}(L)+$ $P(+\|G)P_{(5)}(G)$	0.492	0.455	0.284	0.827	0.788

On the other hand, one can see that the probabilities in data (D) satisfy FTP approximately. The strain of ML308 is a mutant, which can not produce the repressor protein. Therefore, β-galactosidase is always produced in the cell of ML308. ML308-2 is not only *lacI*$^-$ but also *lacY*$^-$, so that the lactose transporting system of ML308-2 is defective and different from that of other strains. Hence the violation of FTP is also seen in data (E).

Here, let us recall Eq. (5.8) in order to calculate lactose/glucose preferences $\{k_L, k_G, \theta\}$. By assigning the values of probabilities $P(+|L)$, $P(+|G)$, $P_{(4)}(L)$, $P_{(5)}(L)$, $P_{(4)}(+)$ and $P_{(5)}(+)$ to Eq. (5.8), we have two equations:

$$\begin{aligned} &P(+|L)\left\{P_{(5)}(L)\Delta - P_{(4)}(L)\Delta'\right\}\frac{k_L}{k_G} + \left\{P_{(4)}(+)\Delta' - P_{(5)}(+)\Delta\right\}\sqrt{\frac{k_L}{k_G}} \\ &+ P(+|G)\left\{P_{(5)}(G)\Delta - P_{(4)}(G)\Delta'\right\} = 0, \end{aligned} \tag{5.9}$$

and

$$\cos\theta = \frac{P_{(4)}(+) - \left\{P(+|L)P_{(4)}(L)\sqrt{\frac{k_L}{k_G}} + P(+|G)P_{(4)}(G)\sqrt{\frac{k_G}{k_L}}\right\}}{2\Delta}, \tag{5.10}$$

where

$$\begin{aligned} \Delta =& P_{(4)}(-)\sqrt{P(+|L)P(+|G)P_{(4)}(L)P_{(4)}(G)} + P_{(4)}(+)\sqrt{P(-|L)P(-|G)P_{(4)}(L)P_{(4)}(G)}, \\ \Delta' =& P_{(5)}(-)\sqrt{P(+|L)P(+|G)P_{(5)}(L)P_{(5)}(G)} + P_{(5)}(+)\sqrt{P(-|L)P(-|G)P_{(5)}(L)P_{(5)}(G)}. \end{aligned}$$

From Eqs. (5.9) and (5.10), we can obtain $\sqrt{k_L/k_G}$ and $\cos\theta$, and Table 5.3 shows these values. We can see that the value of $\sqrt{k_L/k_G}$ is small for data A and C, which violate FTP. On the other hand, the value of $\sqrt{k_L/k_G}$ is nearly one, and $\cos\theta$ is nearly zero in data D, which hold FTP approximately.

We remark that the values of $\left(\sqrt{k_L/k_G}, \cos\theta\right)$ in data A are different from those in data B. This result means that the preculture condition affects *E. coli*'s preference. By assigning the preferences $\left(\sqrt{k_L/k_G}, \theta\right)$ to Eq. (5.8), we compute $P(+)$ with respect to $P(L)$, see Fig. 5.1a. One can see that the behavior of lactose operon depends not only on the strain but on the preculture condition as well; The $P(+)$ of W3110(Gly) in Fig. 5.1a is increasing more linearly than that of W3110(LB). Therefore, the degree of violation of FTP for W3110 precultured on glycerol is less than that for W3110 precultured in LB.

For the data C, D, and E, we plot the values of $P(+)$ with respect to $P(L)$ in Fig. 5.1b. It shows the influence of *lacI* gene or *lacY* gene to lactose operon. The $P(+)$ of ML308, which does not violate FTP so much, is linearly-increasing with respect to $P(L)$.

Table 5.3 *E. coli*'s preferences calculated from each data

	A	B	C	D	E
$\sqrt{\frac{k_L}{k_G}}$	0.066	0.406	0.190	0.838	0.697
$\cos\theta$	−0.842	1.000	−0.461	0.251	−0.733

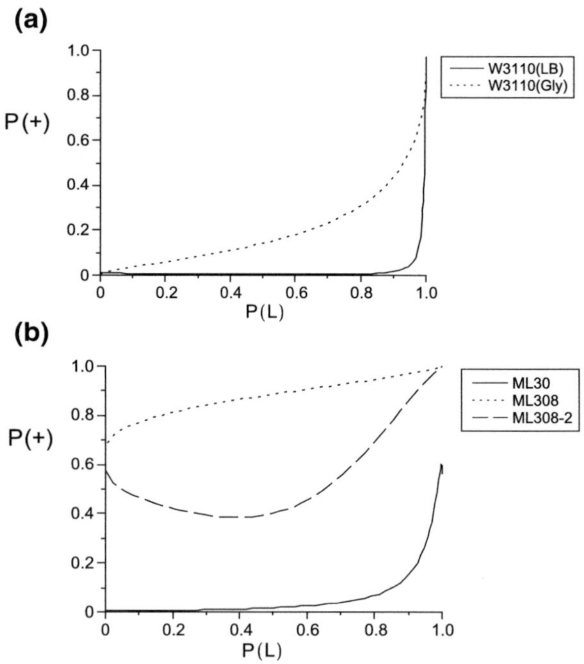

Fig. 5.1 Computed values of $P(+)$ with respect to $P(L)$. **a** Effect of preculture condition. **b** Effect from *lacI* gene and *lacY* gene

References

1. Jacob, F., Monod, J.: J. Mol. Biol. **3**, 318–356 (1961)
2. Inada, T., Kimata, K., Aiba, H.: Mechanism responsible for glucose-lactose diauxie in *Escherichia coli* challenge to the cAMP model. Genes and Cells **1**, 293–301 (1996)
3. Nakao, K., Morita, R., Saji, Y., Ishida, K., Tomita, Y., Ogawa, M., Saitoh, M., Tomooka, Y., Tsuji, T.: The development of a bioengineered organ germ method. Nat. Methods. **4**, 227–230 (2007)
4. Clement, V., Marino, D., Cudalbu, C., Hamou, M.F., Mlynarik, V., de Tribolet, N., Dietrich, P.Y., Gruetter, R., Hegi, M.E., Radovanovic, I.: Marker-independent identification of glioma-initiating (2010)
5. Yam. P.: The Pathological Protein: Mad Cow, Chronic Wasting, and Other Deadly Prion Diseases Springer, Berlin (2003)
6. Miller, J.H.: Experiments in molecular genetics, Cold Spring Harbor Laboratory, Cold Spring Harbor (1972)
7. Bechwith, J.R., Zipser, D.: The Lactose Operon, Cold Spring Harbor Laboratory (1970)

Chapter 6
Application to Decision Making Theory and Cognitive Science

Abstract In this chapter the formalism of quantum probability and quantum information theory (in its generalized form based on theory of lifting) is applied to construct the QL-representations for cognitive processes, especially decision making in games of the prisoner's dilemma type. Our modeling is based on the results of extended studies in the domain of cognitive psychology demonstrated that in some mental contexts players behave irrationally, i.e., they select mixed strategies which are different from the Nash equilibrium predicted by classical game theory. The simplest model of such irrational behavior is based on theory of open quantum systems and quantum master equation. More complex cognitive situations are modelled with aid quantum adaptive dynamics generalizing theory of open quantum systems. We also construct the QL-representation for bistable perception. We are able to construct quantum operators providing the adequate operational description of know statistical data. To check nonclassicality of these data we use a Bell-type inequality, namely, the Leggett-Garg inequality.

Keywords Cognitive processes · Decision making · Games of prisoner's dilemma type · Irrational behavior · Nash equilibrium · Bistable perception · Open quantum systems · Adaptive quantum dynamics · Bell inequality · Leggett-Garg inequality

6.1 Cognitive Psychology and Game Theory

Cognitive psychologists and game theorists have long been aware that it is impossible to use classical (Kolmogorov) probability to describe statistical data collected, e.g., in games of the Prisoners Dilemma type. However, they selected another pathway for the departure from Kolmogorovness, not through probabilistic manifolds with charts of the Kolmogorov type, but through violating the basic axiom of the Kolmogorov's axiomatics of probability theory—additivity of probability. They used a subadditive generalization of a probability measure, see, e.g., [1–4] (see also [5–7] for comparison of nonadditivity and QL approaches). This helped to solve a number of problems of cognitive psychology and game theory. However, the subadditive approach does not have such a degree of generality as the quantum formalism.

M. Asano et al., *Quantum Adaptivity in Biology: From Genetics to Cognition*,
DOI 10.1007/978-94-017-9819-8_6

Subadditive probabilities for different psychological phenomena are constructed, so to say, by hands, without general justification provided by the quantum formalism. Moreover, operating with subadditive probabilities is mathematically more complicated than with linear operators (matrices) in quantum probability theory. Therefore the usage of the non-Kolmogorovean model based on quantum probability is an interesting alternative to the usage of the aforementioned non-Kolmogorovean model based on subadditive probability.

Nowadays the formalism of quantum mechanics is actively applied to game theory and more generally to decision making. The first steps in this direction were done in a series of articles, Conte et al. [8, 9], devoted to the conditional recognition of ambiguous figures, a quantum model for the order effect. The order effect was characterized by the magnitude of the interference term, see Sect. 4.1.1. The idea of representation of the order effect in the form of the QL-interference effect was elaborated in a series of works of Khrennikov [10–12]. However, the real QL-revolution in cognitive psychology started with series of papers of Busemeyer et al. [13, 14] coupling the QL-interference effect with the so called *disjunction effect* (well known in psychology, Hofstader [15, 16], Shafir and Tversky [17, 18], Croson [19]), Savage Sure Thing Principle (the basic principle of modern economic theory formalizing the rationality of decision makers operating at the market, in particular, at the financial market), and order effect.[1]

The works of Busemeyer et al. [13, 14] stimulated the development of QL-schemes of decision making based on the dynamical evolution of the belief-state of a decision maker. The basic model was proposed in the paper [35] of Pothos and Busemeyer and it was based on the combination of the Schrödinger dynamics with the projection type measurement at the last step.

In a series of works of the authors of this book and Basieva [36–41] a novel approach to QL-modeling of decision making was elaborated. It is based on the theory of open quantum systems and more generally adaptive dynamics and the calculus of joint-like probabilities, see Chap. 4. In this chapter we present this approach.

6.2 Decision Making Process in Game

6.2.1 Prisoner's Dilemma

Game theory analyzes the interdependency of decision making entities (players) under a certain institutional condition, and it is applied to various disciplines: social sciences, biology, and philosophy. In a normal game, players have some strategies

[1] Some of these problems were also studied by Cheon and Takahashi [20, 21], Khrennikov [22, 23], Fichtner et al. [24], Franco [25], Khrennikov and Haven [26], Pothos and Busemeyer [27, 28], Lambert-Mogiliansky et al. [29], Dzhafarov and Kujala [30, 31], and Wang et al. [32, 33], see also the special section on cognition, decision making and finances in the conference proceedings [34].

to be chosen and obtain payoffs, which are assigned for the results of their choices. In the player's decision making, the following three assumptions are required:

- Players know the rule of the game; each player knows all selectable strategies and payoffs
- A player behaves rationally so as to maximize his own payoff
- Each rational player recognizes the rationalities of other players.

The goal of game theory is to explain various interdependencies in the real and mental world, and it is believed that the concept of *Nash equilibrium* provides a "rational" solution. However, there are some experiments, in which real players frequently do not achieve a rational solution, and it seems to be "irrational" [16, 17]. There are cases where a real player's decision making process is essentially different from a "rational" player's one in conventional game theory.

We study this problem by using the game given by the payoff table of Table 6.1 as the basic example. This is a well-known two-player game called a prisoner's dilemma (PD) game. The players A and B have two strategies denoted by $0_{A,B}$ and $1_{A,B}$, and the values of payoffs are assigned for the results of $(0_A, 0_B)$, $(0_A, 1_B)$, $(1_A, 0_B)$ and $(1_A, 1_B)$. Generally, PD game has a unique Nash equilibrium: Since the dominant strategies 1_A and 1_B are the best for the rational players A and B, the result $(1_A, 1_B)$ is the Nash equilibrium. A player's decision making process is described as a combination of rational reasoning of one player and her/his estimation of another player as rationally reasoning. The diagram in Fig. 6.1 represent the decision making process of the A-player. The two solid arrows in this diagram express that the player A prefers the result $(1_A, 0_B)$ to $(0_A, 0_B)$ and prefers $(1_A, 1_B)$ to $(0_A, 1_B)$. The two dotted arrows explains that the player A reasons that the player B will prefer $(0_A, 1_B)$ to $(0_A, 0_B)$ and $(1_A, 1_B)$ to $(1_A, 0_B)$. These arrows represent A's preference. One can see that A's reasoning always leads to the result $(1_A, 1_B)$, namely, the player A chooses 1_A believing that the player B will choose 1_B. However, we wonder whether such a description of the decision making process provides the complete picture of reasoning of a real player.

Table 6.1 An example of payoff table in prisoner's dilemma

A/B	0_B	1_B
0_A	$4/4$	$2/5$
1_A	$5/2$	$3/3$

Fig. 6.1 Player's preferences

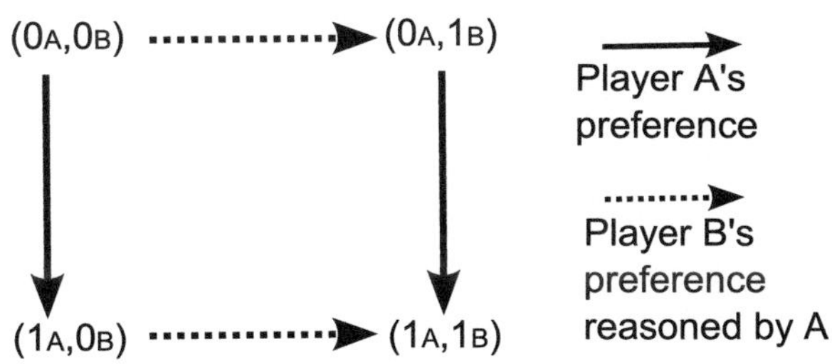

Table 6.2 Extreamly high payoffs

A/B	0_B	1_B
0_A	100000/100000	2/100000 + 1
1_A	100000 + 1/ 2	3/3

Let us consider another PD game represented by the payoff matrix of Table 6.2; In this game, a real player will choose the strategy 0, because such a player may assume that the payoff 100000 for the selection $(0_A, 0_B)$ of the strategies is very attractive compared with the selection $(1_A, 1_B)$, and this choice is also attractive for the other player. Moreover, she/he may reason "that the player B will think in the same way." She/He cannot neglect this possibility in her/his decision making, and this opens the possibility to shift the expectation from $(1_A, 1_B)$ to $(0_A, 0_B)$. We express such a shift as the diagonal arrows in the diagram of Fig. 6.2.

In this situation, player A's reasoning would be cyclic such as

$$(0_A, 0_B) \longrightarrow (1_A, 0_B) \longrightarrow (1_A, 1_B) \longrightarrow (0_A, 0_B) \longrightarrow \cdots$$

Then her/his choices would oscillate between the strategies 1_A and 0_A. However, it is clear that the intention to choose 0_A is dominant since it comes from the payoff-difference between 100000 of $(0_A, 0_B)$ and 3 of $(1_A, 1_B)$.

The additional shift might seem to be strange. In the framework of game theory, the player A's preferences are explained in the following way. "If the player B chosen $x_B \in \{0_B, 1_B\}$, then I would prefer to choose $y_A \in \{0_A, 1_A\}$", and the player B's preferences reasoned by A are explained in the same form. Such a mental process consists of two parts; the part of *assumption* and the part of *analysis*. In the first part, the player A assumes one of the player B's choices. In the second part, the player A analyzes his preference based on the assumption. Note that such an analysis is contextually the same as the *posterior analysis* of someone who already knows the result of B's choice. All of the shifts seen in Fig. 6.1 are provided by such posterior analysis. On the other hand, the additional shift in Fig. 6.2 is *never* induced by the posterior analysis, since the player B's choice is not fixed on this shift.

In principle, each player is not informed about another player's choice, that is, each player feels *uncertainty* about another player's choice. A *prior analysis* should be done in the presence of uncertainty. The prior analysis in the conventional

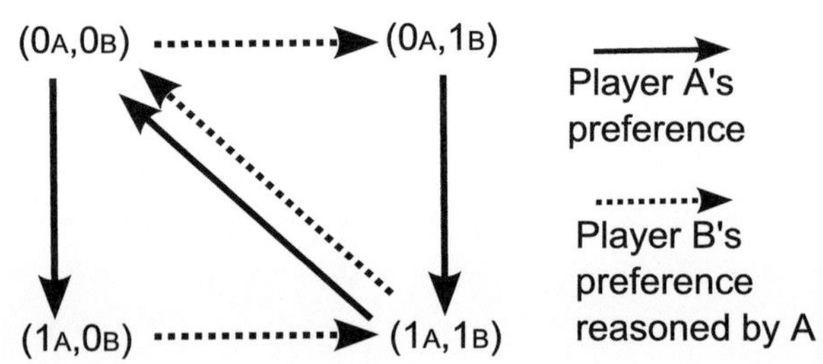

Fig. 6.2 Diagonal preferences

game theory is just a *mixture of the posterior analysis* based on a *mixed strategy*, a probabilistic distribution given for another player's choice: The "rational" player A will consider the posterior analysis of the possibilities for two strategies mixed with probabilities $P(0_B)$ and $P(1_B)$. In each posterior analysis, the player A can estimate probabilities of his own choices, $P(i_A|0_B)$, $(i = 0, 1)$ and $P(i_A|1_B)$, $(i = 0, 1)$. For the mixed strategy, the player A makes his own choice with the probability given by FTP:

$$P(i_A) = P(i_A|0_B)P(0_B) + P(i_A|1_B)P(1_B).$$

For example, in PD game the $P(1_A)$ always becomes 1. As was mentioned above, the additional shift in Fig. 6.2 is not induced by posterior analysis. Therefore, if the additional shift affects the decision making, the total probability law will be violated;

$$P(i_A) \neq P(i_A|0_B)P(0_B) + P(i_A|1_B)P(1_B).$$

Actually, such non-Kolmogorovian feature can be seen in the statistical data obtained in some experiments, see [16, 17]. The additional effect comes from a "deeper" uncertainty, where representation of possible actions of another player fluctuates. Such mental fluctuation cannot be represented in terms of classical probability theory.

Quantum probability is one of the most advanced mathematical models for non-classical probability, and it is useful to represent the psychological uncertainty, which a real player feels. In this section, we introduce "a mental state", which represents mental fluctuations modelled as quantum fluctuations and describe the process of decision making as the dynamic of the mental state. This dynamics is also represented in the quantum framework.

6.2.2 Mental State and Its Dynamics

By following the tradition of quantum information theory we call the A-player Alice and the B-player Bob. We focus on Alice's reasoning. In principle, Alice does not know which action Bob will choose, that is, she feels uncertainty about Bob's action. We represent such uncertainty with aid of the following state vector:

$$|\phi_B\rangle = \alpha|0_B\rangle + \beta|1_B\rangle \in \mathcal{H}_B = \mathbb{C}^2 \cdot \ (|\alpha|^2 + |\beta|^2 = 1) \tag{6.1}$$

The orthogonal basis $|0_B\rangle$ and $|0_B\rangle$ corresponds to Alice's judgings "Bob will choose the action 0" and "Bob will choose the action 1". The structure of the state of $|\phi_B\rangle$ implies that Alice essentially cannot deny either of the two possibilities. The values of α and β determine the weights assigned to these possibilities. We call the state $|\phi_B\rangle\langle\phi_B| \equiv \sigma_B$ "a prediction state". (Here we represented the pure state (6.1), a vector in complex Hilbert space, by the corresponding density operator, see Sect. 2.2. In our coming modeling it is better to proceed with density operators. The process of the decision making will be described by the quantum master equation which transfers pure states given by vectors into in general mixed states given by density operators.)

Alice will be aware of two consequences of "$0_A 0_B$" and "$0_A 1_B$" with the weights (probability amplitudes) of α and β, for her own action "0_A". This situation is represented by the vector

$$\begin{aligned} |\Phi_{0_A}\rangle &= \alpha|0_A 0_B\rangle + \beta|0_A 1_B\rangle \\ &= |0_A\rangle \otimes |\phi_B\rangle \in \mathscr{H}_A \otimes \mathscr{H}_B = \mathbb{C}^2 \otimes \mathbb{C}^2 \end{aligned} \tag{6.2}$$

Similarly,

$$|\Phi_{1_A}\rangle = |1_A\rangle \otimes |\phi_B\rangle, \tag{6.3}$$

is given for the action "1_A". We call $|\Phi_{0_A}\rangle$ and $|\Phi_{1_A}\rangle$ "alternative state vectors". In more general forms, the alternative state vectors are given as

$$\begin{aligned} |\Phi_{0_A}\rangle &= |0_A\rangle \otimes |\phi_B^0\rangle, \\ |\Phi_{1_A}\rangle &= |1_A\rangle \otimes |\phi_B^1\rangle, \end{aligned} \tag{6.4}$$

with different prediction state vectors,

$$\begin{aligned} |\phi_B^0\rangle &= \alpha_0|0_B\rangle + \beta_0|1_B\rangle, \\ |\phi_B^1\rangle &= \alpha_1|0_B\rangle + \beta_1|1_B\rangle. \ (|\alpha_i|^2 + |\beta_i|^2 = 1) \end{aligned} \tag{6.5}$$

By using the alternative state vectors, we consider the state defined as

$$\sum_{i,j=0,1} x_{ij}|\Phi_{i_A}\rangle\langle\Phi_{j_A}| \equiv \Theta. \tag{6.6}$$

This state always satisfies the normalization condition:

$$\sum_{i=0,1} \mathrm{tr}(\Theta|\Phi_{i_A}\rangle\langle\Phi_{i_A}|) = 1, \tag{6.7}$$

and it specifies Alice's mental state such that her action 0_A (or 1_A) is decided with the probability $\mathrm{tr}(\Theta|\Phi_{0_A}\rangle\langle\Phi_{0_A}|) \equiv P_{0_A}$ (or $\mathrm{tr}(\Theta|\Phi_{1_A}\rangle\langle\Phi_{1_A}|) \equiv P_{1_A}$). We call Θ the "mental state".

The process of decision making in a game is described as the dynamics of the mental state Θ.

Initial Mental State

Before the decision making, Alice's mental state $\Theta(t = 0) = \Theta_0$ at the time $t = 0$ is given as

$$\Theta_0 = |\Phi(0)\rangle\langle\Phi(0)|, \tag{6.8}$$

where

$$|\Phi(0)\rangle = x_0|\Phi_{0A}\rangle + x_1|\Phi_{1A}\rangle.$$

x_0 and x_1 are complex numbers with $|x_0|^2 + |x_1|^2 = 1$. The values of x_0 and x_1 are not very important because Alice has not started her thinking for decision making yet. In general, they are determined randomly with the normalization. To simplify the discussion, we assume $x_0 = 1$, that is, $\Theta_0 = |\Phi_{0A}\rangle\langle\Phi_{0A}|$.

Alice's Decision Making Dynamics

Alice will make her decision by thinking about the game given. In other words, the dynamics of the mental state is based on Alice's self-reflections. In our model it is assumed that Alice can reason in two following ways:

Resoning ($0 \to 1$): "The choice of 1 is better than the choice of 0."
Resoning ($1 \to 0$): "The choice of 0 is better than the choice of 1."

These reasonings decreases Alice's tendency for the choice of 0 (or 1) and motivates her choice of 1 (or 0). For such a shift of Alice's mental state, the following assumptions are supposed to be valid:

1. The shift is occurred in a certain time interval τ.
2. The shift is *repeated* until the mental state becomes stable.
3. All of the repeated shifts are *memorized* by Alice.

Memory Space and Branching Map

We represent the shift induced by reasoning ($0 \to 1$) and ($1 \to 0$) with the aid of the diagram of "branching" as Fig. 6.3.

Figure 6.3 shows the branching of the intention to the choice of 0. In accordance with the third assumption above, Alice memorizes these two paths. Here, the new

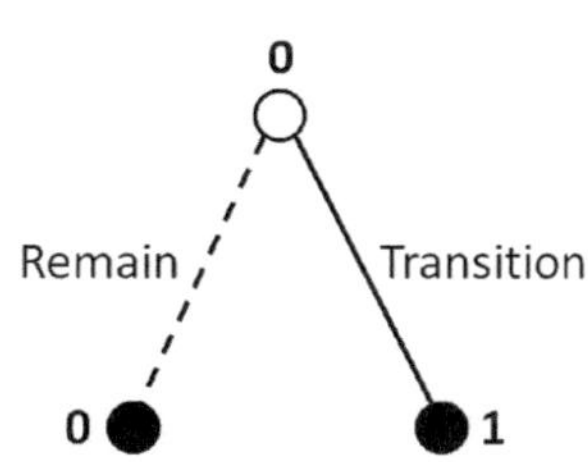

Fig. 6.3 The branching of the intention to the choice of 0: The intention to the choice of 0 is partially transited to the choice of 1 (the *solid path*) and partially it is preserved (the *dotted path*)

space $\mathcal{M} = \mathbb{C}^2$ called "memory space" is introduced, and the above branching is described as a vector from the space $\mathcal{H}_A \otimes \mathcal{H}_B \otimes \mathcal{M}$;

$$\begin{aligned} |\Phi_{0\to 1}\rangle &= \sqrt{1-\Delta_\tau}|\Phi_{0_A}\rangle \otimes |R\rangle + \sqrt{\Delta_\tau}|\Phi_{1_A}\rangle \otimes |T\rangle \\ &= \sqrt{1-\Delta_\tau}|\Phi_{0_A} R\rangle + \sqrt{\Delta_\tau}|\Phi_{1_A} T\rangle. \end{aligned}$$

$\{|R\rangle, |T\rangle\}$ is the orthonormal basis in the space $\mathcal{M}$, and $|R\rangle, |T\rangle$ represent the dotted path and solid path, respectively. The value of Δ_τ $(0 \le \Delta_\tau \le 1)$ is the degree of intensity to shift from 0 to 1. ($\sqrt{\Delta_\tau}$ is the complex number satisfying $|\sqrt{\Delta_\tau}|^2 = \Delta_\tau$.) In a similar way, the vector $|\Phi_{1\to 0}\rangle$, which represents the branching to choice of 1, is written as

$$\begin{aligned} |\Phi_{1\to 0}\rangle &= \sqrt{1-\tilde{\Delta}_\tau}|\Phi_{1_A}\rangle \otimes |R\rangle + \sqrt{\tilde{\Delta}_\tau}|\Phi_{0_A}\rangle \otimes |T\rangle \\ &= \sqrt{1-\tilde{\Delta}_\tau}|\Phi_{1_A} R\rangle + \sqrt{\tilde{\Delta}_\tau}|\Phi_{0_A} T\rangle \end{aligned}$$

The parameter $\sqrt{\tilde{\Delta}_\tau}$ specifies the degree of intensity to shift from 1 to 0. It is assumed that such branching is occurred in the time interval τ, so Δ_τ and $\tilde{\Delta}_\tau$ are functions of τ in general.

As the next step, the "branching map" $Z_\tau : \mathcal{H}_A \otimes \mathcal{H}_B \mapsto \mathcal{H}_A \otimes \mathcal{H}_B \otimes \mathcal{M}$ is introduced. This map generates the state change of branching;

$$\begin{aligned} Z_\tau|\Phi_{0_A}\rangle &= |\Phi_{0\to 1}\rangle, \\ Z_\tau|\Phi_{1_A}\rangle &= |\Phi_{1\to 0}\rangle. \end{aligned} \tag{6.9}$$

The branching map Z_τ is decomposed as $V_\tau \circ M$. The map $M : \mathcal{H}_A \otimes \mathcal{H}_B \mapsto \mathcal{H}_A \otimes \mathcal{H}_B \otimes \mathcal{M}$ plays the role of "memory" generator;

$$\begin{aligned} M|\Phi_{0_A}\rangle &= |\Phi_{0_A} R\rangle, \\ M|\Phi_{1_A}\rangle &= |\Phi_{1_A} R\rangle, \end{aligned} \tag{6.10}$$

and $V_\tau : \mathcal{H}_A \otimes \mathcal{H}_B \otimes \mathcal{M} \mapsto \mathcal{H}_A \otimes \mathcal{H}_B \otimes \mathcal{M}$ is a unitary operator generating the shift;

$$\begin{aligned} V_\tau|\Phi_{0_A} R\rangle &= |\Phi_{0\to 1}\rangle, \\ V_\tau|\Phi_{1_A} R\rangle &= |\Phi_{1\to 0}\rangle. \end{aligned} \tag{6.11}$$

The unitary operator V_τ has the form of time evolution operator;

$$V_\tau = \mathrm{e}^{-\mathrm{i}\lambda H \tau}. \tag{6.12}$$

The operator H is the Hamiltonian of the interaction between the "operative mental state" and the state of the memory. It represents the generator of Alice's decision making. In Sect. 7.1.4, we shall discuss the construction of H in detail. The parameter λ in $e^{-iH\lambda}$ determines the strength of interaction.

Dynamics Represented by Lifting Map

As the result of action of the branching map Z_τ, the initial state $\Theta(0) = \Theta_0$ is mapped into the state belonging on $\mathcal{H}_A \otimes \mathcal{H}_B \otimes \mathcal{M}$;

$$\Theta_0 \mapsto Z_\tau \Theta_0 Z_\tau^*$$

The new mental state at $t = \tau$ is given by

$$\mathrm{tr}_{\mathcal{M}}(Z_\tau \Theta_0 Z_\tau^*) \equiv \Theta(\tau) = \Theta_1.$$

As was mentioned in the third assumption, in the process of decision making, this shift is repeated. In other words, the application of Z_τ is repeated. In the next time interval τ, the following state is produced:

$$(Z_\tau \otimes I) Z_\tau \Theta_0 Z_\tau^* (Z_\tau^* \otimes I) \in \mathcal{S}(\mathcal{H}_A \otimes \mathcal{H}_B \otimes \mathcal{M} \otimes \mathcal{M}).$$

Here, we note that the new memory space $\mathcal{M}$ is added for memorizing the paths of the new branching. When the initial mental state is given by $\Theta_0 = |\Phi_{0_A}\rangle\langle\Phi_{0_A}|$, the following mental state is produced:

$$\begin{aligned}
&(Z_\tau \otimes I) Z_\tau \Theta_0 Z_\tau^* (Z_\tau^* \otimes I) = |\Phi(2\tau)\rangle\langle\Phi(2\tau)|,\\
&|\Phi(2\tau)\rangle = (1-\Delta)|\Phi_{0_A}\rangle \otimes |RR\rangle + \sqrt{\Delta\tilde{\Delta}}|\Phi_{0_A}\rangle \otimes |TT\rangle\\
&\quad + \sqrt{\Delta(1-\tilde{\Delta})}|\Phi_{1_A}\rangle \otimes |TR\rangle + \sqrt{\Delta(1-\Delta)}|\Phi_{1_A}\rangle \otimes |RT\rangle.
\end{aligned}$$

In the above expression, $|RR\rangle$, $|TT\rangle$, $|TR\rangle$ and $|RT\rangle$ belonging on $\mathcal{M} \otimes \mathcal{M}$ represent all the possible paths created as the result of branching, see Fig. 6.4. The mental state at $t = 2\tau$ is given by

$$\mathrm{tr}_{\mathcal{M}\otimes\mathcal{M}}((Z_\tau \otimes I) Z_\tau \Theta_0 Z_\tau^* (Z_\tau^* \otimes I)) = \mathrm{tr}_{\mathcal{M}}(Z_\tau \Theta_1 Z_\tau^*) \equiv \Theta(2\tau) = \Theta_2.$$

The state $\Theta(n\tau) = \Theta_n$ obtained after n shifts is described as

$$\Theta_n = \mathrm{tr}_{\mathcal{M}_n \otimes \cdots \otimes \mathcal{M}_1}(\mathcal{E}_n^*(\mathcal{E}_{n-1}^*(\cdots \mathcal{E}_2^*(\mathcal{E}_1^*(\Theta_0))\cdots))). \tag{6.13}$$

Here, $\mathcal{E}_j^*$ is the lifting map [42] from $B(\mathcal{H}_A \otimes \mathcal{H}_B \otimes \mathcal{M}_{j-1} \otimes \cdots \otimes \mathcal{M}_2 \otimes \mathcal{M}_1)$ to $B(\mathcal{H}_A \otimes \mathcal{H}_B \otimes \mathcal{M}_j \otimes \mathcal{M}_{j-1} \otimes \cdots \otimes \mathcal{M}_2 \otimes \mathcal{M}_1)$ that is defined by

$$\mathcal{E}_j^*(\cdot) = (Z \otimes I \otimes \cdots \otimes I) \cdot (Z^* \otimes I \otimes \cdots \otimes I). \tag{6.14}$$

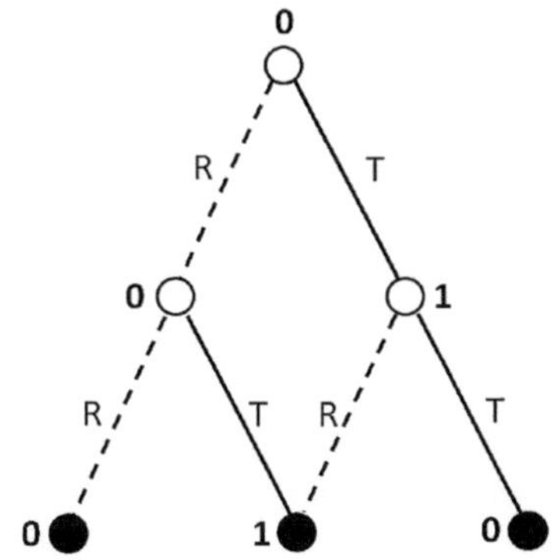

Fig. 6.4 The branchings at $t = 2\tau$: The intention to the choice of 0 is partially transited to the choice of 1 (the *solid path*) and partially it is preserved (the *dotted path*)

Stabilization of Mental State

The mental state $\Theta_n = \sum_{i,j=0,1} x_{ij}(n)|\Phi_{i_A}\rangle\langle\Phi_{j_A}|$ can be expanded with respect to the basis $|\Psi_{0_A}\rangle$ and $|\Psi_{1_A}\rangle$, and it can be represented as 2×2 matrix;

$$\Theta_n = \begin{pmatrix} x_{00}(n) & x_{01}(n) \\ x_{10}(n) & x_{11}(n) \end{pmatrix}.$$

The diagonal parts x_{00} and x_{11} correspond to the probabilities of choices P_{0_A} and P_{1_A}. From Eq. (6.13) we can obtain the following results:

$$\begin{aligned} P_{0_A}(n) &= (1 - \Delta_\tau - \tilde{\Delta}_\tau)^n P_{0_A}(0) + (1 - (1 - \Delta_\tau - \tilde{\Delta}_\tau)^n) \frac{\tilde{\Delta}_\tau}{\Delta_\tau + \tilde{\Delta}_\tau} \\ &= (1 - \xi)^n P_{0_A}(0) + (1 - (1 - \xi)^n) P^S_{0_A}, \\ P_{1_A}(n) &= (1 - \Delta_\tau - \tilde{\Delta}_\tau)^n P_{1_A}(0) + (1 - (1 - \Delta_\tau - \tilde{\Delta}_\tau)^n) \frac{\Delta_\tau}{\Delta_\tau + \tilde{\Delta}_\tau} \\ &= (1 - \xi)^n P_{1_A}(0) + (1 - (1 - \xi)^n) P^S_{1_A}. \end{aligned} \tag{6.15}$$

Here, $\Delta_\tau + \tilde{\Delta}_\tau = \xi$, $\frac{\tilde{\Delta}_\tau}{\Delta_\tau + \tilde{\Delta}_\tau} = P^S_{0_A}$ and $\frac{\Delta_\tau}{\Delta_\tau + \tilde{\Delta}_\tau} = P^S_{1_A}$. Noting that $0 \leq \xi \leq 2$, we can find that the probabilities $P_{0_A}(n)$ and $P_{1_A}(n)$ approach the stable values $P^S_{0_A}$ and $P^S_{1_A}$ after a large numbers n of iterations. We also note that the non-diagonal parts $x_{01}(n)$ and $x_{10}(n)$ approach zero. Thus in this model in the process of decision making the mental state is stabilized to the state represented by the density matrix:

$$\Theta_E = \begin{pmatrix} P^S_{0_A} & 0 \\ 0 & P^S_{1_A} \end{pmatrix}.$$

Construction of Hamiltonian H

In Eq. (6.11), we introduced the time evolution operator V_τ that plays a main role of the structure of the branching operator Z_τ. Alice's intentions for reasonings $(0 \to 1)$ and $(1 \to 0)$ is encoded in this operator V_τ. In Eq. (6.12), we defined V_τ as the time evolution operator $e^{i\lambda H\tau}$. In this section, the corresponding Hamiltonian H is constructed by using an operator that specifies Alice's so to say "decision making processor".

Alice has to *compare* all potential consequences of the game

$$\{0_A 0_B, 0_A 1_B, 1_A 0_B, 1_A 1_B\}.$$

We introduce the operator $G \in \mathcal{H}_B(\mathcal{H}_A \otimes \mathcal{H}_B)$ which is represented as the matrix

$$G = \begin{pmatrix} 0 & 0 & \tilde{\mu}_1 & \tilde{\mu}_3 \\ 0 & 0 & \tilde{\mu}_2 & \tilde{\mu}_4 \\ \mu_1 & \mu_2 & 0 & 0 \\ \mu_3 & \mu_4 & 0 & 0 \end{pmatrix}$$

in the basis $\{|0_A 0_B\rangle, |0_A 1_B\rangle, |1_A 0_B\rangle, |1_A 1_B\rangle\}$. The elements $\{\mu_i\}$ and $\{\tilde{\mu}_j\}$ are real numbers. We call this operator the "comparison operator". For example, when Alice has the tendency to prefer the consequence $1_A 0_B$ to $0_A 0_B$, its degree of preference is given by the value of $\mu_1 = \langle 1_A 0_B | G | 0_A 0_B \rangle$. The degree of the opposite tendency is given by the quantity $\tilde{\mu}_1 = \langle 0_A 0_B | G | 1_A 0_B \rangle$. (In general, it is allowed that these two tendencies coexist in Alice's mental representation of the game.)

Here, we recall that Alice is not informed about Bob's choice. Alice's intention to choose 0_A always fluctuates between $0_A 0_B$ and $0_A 1_B$. This is a consequence of the form of the alternative vector $|\Phi_{0_A}\rangle = \alpha_0 |0_A 0_B\rangle + \beta_0 |0_A 1_B\rangle$. Similarly, Alice with the mind to choose 1_A is fluctuated between $1_A 0_B$ and $1_A 1_B$, as seen in $|\Phi_{1_A}\rangle = \alpha_1 |1_A 0_B\rangle + \beta_1 |1_A 1_B\rangle$. We consider the values of $\langle \Phi_{1_A} | G | \Phi_{0_A} \rangle$ and $\langle \Phi_{0_A} | G | \Phi_{1_A} \rangle$, which are calculated as

$$\begin{aligned} \langle \Phi_{1_A} | G | \Phi_{0_A} \rangle &= \alpha_0 \alpha_1^* \mu_1 + \beta_0 \alpha_1^* \mu_2 + \alpha_0 \beta_1^* \mu_3 + \beta_0 \beta_1^* \mu_4 \equiv \mu, \\ \langle \Phi_{0_A} | G | \Phi_{1_A} \rangle &= \alpha_0^* \alpha_1 \tilde{\mu}_1 + \beta_0^* \alpha_1 \tilde{\mu}_2 + \alpha_0^* \beta_1 \tilde{\mu}_3 + \beta_0^* \beta_1 \tilde{\mu}_4 \equiv \tilde{\mu}. \end{aligned} \tag{6.16}$$

These values μ and $\tilde{\mu}$ specify the degrees intentions for transitions $(0 \to 1)$ and $(1 \to 0)$.

By using the comparison operator, we define the Hamiltonian H in Eq. (6.12);

$$\begin{aligned} H = &\left|\Phi_{1_A} T\right\rangle\left\langle\Phi_{1_A} T\right| K \left|\Phi_{0_A} R\right\rangle\left\langle\Phi_{0_A} R\right| \\ &+ \left|\Phi_{0_A} T\right\rangle\left\langle\Phi_{0_A} T\right| K \left|\Phi_{1_A} R\right\rangle\left\langle\Phi_{1_A} R\right| \\ &+ \left|\Phi_{0_A} R\right\rangle\left\langle\Phi_{0_A} R\right| K^* \left|\Phi_{1_A} T\right\rangle\left\langle\Phi_{1_A} T\right| \\ &+ \left|\Phi_{1_A} R\right\rangle\left\langle\Phi_{1_A} R\right| K^* \left|\Phi_{0_A} T\right\rangle\left\langle\Phi_{0_A} T\right|, \end{aligned} \tag{6.17}$$

where

$$K = G \otimes |T\rangle\langle R| .$$

Equation (6.16) implies that this H is rewritten as

$$\begin{aligned} H = &\mu \left|\Phi_{1_A} T\right\rangle\left\langle \Phi_{0_A} R\right| + \tilde{\mu} \left|\Phi_{0_A} T\right\rangle\left\langle \Phi_{1_A} R\right| \\ &+ \mu^* \left|\Phi_{0_A} R\right\rangle\left\langle \Phi_{1_A} T\right| + \tilde{\mu}^* \left|\Phi_{1_A} R\right\rangle\left\langle \Phi_{0_A} T\right| . \end{aligned} \tag{6.18}$$

One can check that the vectors

$$|\Psi_{\pm}\rangle = \frac{1}{\sqrt{2}} \left|\Phi_{0_A} R\right\rangle \pm \frac{e^{i\theta}}{\sqrt{2}} \left|\Phi_{1_A} T\right\rangle$$

are eigenvectors of H for eigenvalues $\pm|\mu| = \pm\omega$ ($e^{i\theta}$ is equal to $\frac{\mu}{|\mu|}$), and the vectors

$$\left|\Psi'_{\pm}\right\rangle = \frac{1}{\sqrt{2}} \left|\Phi_{1_A} R\right\rangle \pm \frac{e^{i\tilde{\theta}}}{\sqrt{2}} \left|\Phi_{0_A} T\right\rangle$$

are eigenvectors for eigenvalues $\pm|\tilde{\mu}| = \pm\tilde{\omega}$ ($e^{i\tilde{\theta}} = \frac{\tilde{\mu}}{|\tilde{\mu}|}$). Further, one can check that the time evolution operator $V_\tau = e^{-i\lambda H\tau}$ transforms $\left|\Phi_{0_A} R\right\rangle$ and $\left|\Phi_{1_A} R\right\rangle$ as

$$\begin{aligned} V_\tau \left|\Phi_{0_A} R\right\rangle &= \cos(\omega\lambda\tau) \left|\Phi_{0_A} R\right\rangle - ie^{i\theta} \sin(\omega\lambda\tau) \left|\Phi_{1_A} T\right\rangle, \\ V_\tau \left|\Phi_{1_A} R\right\rangle &= \cos(\tilde{\omega}\lambda\tau) \left|\Phi_{1_A} R\right\rangle - ie^{i\tilde{\theta}} \sin(\tilde{\omega}\lambda\tau) \left|\Phi_{0_A} T\right\rangle . \end{aligned} \tag{6.19}$$

Here, the relations $\left|\Phi_{0_A} R\right\rangle = \frac{1}{\sqrt{2}} |\Psi_+\rangle + \frac{1}{\sqrt{2}} |\Psi_-\rangle$ and $\left|\Phi_{1_A} R\right\rangle = \frac{1}{\sqrt{2}} \left|\Psi'_+\right\rangle + \frac{1}{\sqrt{2}} \left|\Psi'_-\right\rangle$ are used. Since the above results correspond to $V_\tau \left|\Phi_{0_A} R\right\rangle = |\Phi_{0\to1}\rangle$ and $V_\tau \left|\Phi_{1_A}, R\right\rangle = |\Phi_{1\to0}\rangle$, we give $|\sqrt{\Delta_\tau}|$ and $|\sqrt{\tilde{\Delta}_\tau}|$ by

$$|\sqrt{\Delta_\tau}| = \sin(\omega\lambda\tau),\ |\sqrt{\tilde{\Delta}_\tau}| = \sin(\tilde{\omega}\lambda\tau).$$

6.2.3 Numerical Analysis of Dynamics

Setting of Parameters

In this section, numerical analyses of the dynamics of decision making in PD games are presented. Let us consider a two-player game with the following pay-off table.

A/B	0_B	1_B
0_A	a/a'	b/b'
1_A	c/c'	d/d'

The parameters μ_i and $\tilde{\mu}_i$ represent the degrees of Alice's preferences in the comparison of consequences. Here, it is assumed that the values of $(\mu_i^2, \tilde{\mu}_i^2) \equiv (k_i, \tilde{k}_i)$ are determined by the differences of pay-offs: For the comparisons

$$0_A 0_B \underset{\tilde{\mu}_1}{\overset{\mu_1}{\rightleftharpoons}} 1_A 0_B, \; 0_A 1_B \underset{\tilde{\mu}_4}{\overset{\mu_4}{\rightleftharpoons}} 1_A 1_B,$$

$$(k_1, \tilde{k}_1) = \begin{cases} (c-a, 0) \text{ if } a < c \\ (0, a-c) \text{ if } a > c \end{cases},$$
$$(k_4, \tilde{k}_4) = \begin{cases} (d-b, 0) \text{ if } b < d \\ (0, b-d) \text{ if } b > d \end{cases}. \tag{6.20}$$

For the other two comparisons

$$0_A 1_B \underset{\tilde{\mu}_2}{\overset{\mu_2}{\rightleftharpoons}} 1_A 0_B, \; 0_A 0_B \underset{\tilde{\mu}_3}{\overset{\mu_3}{\rightleftharpoons}} 1_A 1_B,$$

$$(k_2, \tilde{k}_2) = \begin{cases} (\sqrt{(c-b)(c'-b')}, 0) \text{ if } b < c \text{ and } b' < c' \\ (0, \sqrt{(b-c)(b'-c')}) \text{ if } b > c \text{ and } b' > c' \\ (0, 0) \text{ in other cases} \end{cases}$$
$$(k_3, \tilde{k}_3) = \begin{cases} (\sqrt{(d-a)(d'-a')}, 0) \text{ if } a < d \text{ and } a' < d \\ (0, \sqrt{(a-d)(a'-d')}) \text{ if } a > d \text{ and } a' > d' \\ (0, 0) \text{ in other cases} \end{cases}. \tag{6.21}$$

In the case of PD game ($c > a > d > b$ and $b' > a' > d' > c'$),

$$\begin{aligned} & k_1 = c - a, \; \tilde{k}_3 = \sqrt{(a-d)(a'-d')} \\ & k_4 = d - b, \\ & \tilde{k}_1 = k_2 = \tilde{k}_2 = k_3 = \tilde{k}_4 = 0. \end{aligned} \tag{6.22}$$

Analysis of the PD Game

First, let us consider the values of $(k_i, \tilde{k}_i)$ in the following PD game, and calculate the probabilities of "irrationality" P_{0_A}. Figure 6.5 shows P_{0_A} for the four values of the parameters; $(\alpha_0^2, \alpha_1^2) = (1 - \beta_0^2, 1 - \beta_1^2) = (0, 0)$, $(0.5, 0.5)$, $(0.75, 0.25)$ and $(1, 0)$. Here, $\alpha_{0,1}$ ($\beta_{0,1}$) are assumed to be real. The strength of the interaction λ and

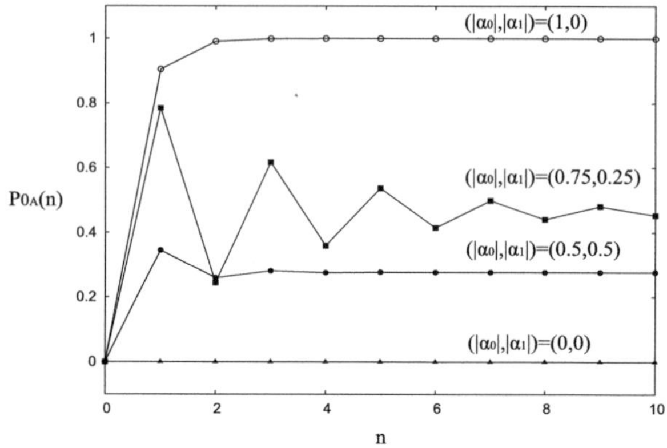

Fig. 6.5 The irrational probabilities in PD game I

the time interval τ are set as $\lambda = \frac{1}{max(\omega,\tilde{\omega})}$ and $\tau = \frac{4\pi}{5}$. It is assumed that before the decision making, at $n = 0$, Alice does not feel the possibility of the irrational choice "0", that is, $P_{0_A}(n = 0) = 0$. At the condition $(\alpha_0^2, \alpha_1^2) = (0, 0)$, Alice judges that "Bob will choose the rational 1." She will not feel uncertainty about Bob's choice and choose "1" since the pay-off 3 is larger than 2. At the condition $(\alpha_0^2, \alpha_1^2) = (1, 0)$, Alice judges that "Bob's strategy will be the same as mine." Then, Alice will choose "0" by comparing the pay-offs 4 and 3. As seen in Fig. 6.5, the probability P_{0_A} develops and approaches 1. At the condition parametrized by $(\alpha_0^2, \alpha_1^2) = (0.5, 0.5)$ or $(0.75, 0.25)$, Alice feels uncertainty and chooses 0 with $0 < P_{0_A} < 1$. In the above calculation, as n becomes large, the probability P_{0_A} approaches

$$P_{0_A}^E = \lim_{n\to\infty} P_{0_A}(n) = \begin{cases} \dfrac{\sin^2(\frac{4\pi}{5}\frac{\tilde{\omega}}{\omega})}{\sin^2(\frac{4\pi}{5})+\sin^2(\frac{4\pi}{5}\frac{\tilde{\omega}}{\omega})} & (\text{if } \omega \geq \tilde{\omega}) \\ \dfrac{\sin^2(\frac{4\pi}{5})}{\sin^2(\frac{4\pi}{5}\frac{\omega}{\tilde{\omega}})+\sin^2(\frac{4\pi}{5})} & (\text{if } \tilde{\omega} \geq \omega) \end{cases}.$$

When $\omega = \tilde{\omega}$, the probability $P_{0_A}^E$ is $\frac{1}{2}$ and the parameters α_0 and, α_1 satisfy

$$\alpha_1 = \sin\left(\arctan\left(1 - \frac{\sqrt{1-\alpha_0^2}}{\alpha_0}\right)\right).$$

In a sense, this function gives a boundary between rational solutions ($P_{0_A}^E < 1/2$) and irrational solutions ($P_{0_A}^E > 1/2$), see Fig. 6.6.

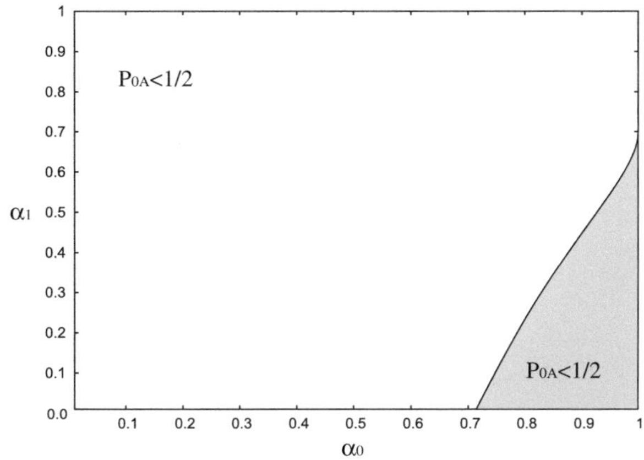

Fig. 6.6 The boundary between rational solutions and irrational solutions in PD game I

If Alice makes her decision with (α_0, α_1) on the boundary, she may strongly feel a "dilemma": See Fig. 6.7, which shows $P_{0_A}(n)$ at $n = 1, 2, 3, 4, 7, 8$ and 19. One can see that the probability $P_{0_A}(n)$ is intensively fluctuated near the boundary.

The irrational choice in PD comes from the fact that the pay-off assigned for the consequence $0_A 0_B$ is larger than the one for $1_A 1_B$. In the case of the game I, the difference of pay-offs of these consequences is only $4 - 3 = 1$. Therefore, as can be seen from Fig. 6.6, the domain of irrational solutions is relatively small. In Fig. 6.8, we introduce the two games denoted by game II and game III in which pay-offs assigned for $0_A 0_B$ are 6 and 10. The boundaries in Fig. 6.8 are given by

$$\text{II: } \alpha_1 = \sin\left(\arctan\left(3 - \frac{\sqrt{1-\alpha_0^2}}{\alpha_0}\right)\right),$$

$$\text{III: } \alpha_1 = \sin\left(\arctan\left(10 - \frac{\sqrt{1-\alpha_0^2}}{\alpha_0}\right)\right),$$

and these domains are larger than those in the game I. The PD game discussed in introduction to this chapter is an extreme example with a large domain. It is expected that in the above game,

A/B	0_B	1_B
0_A	$100000/100000$	$2\ /100000 + 1$
1_A	$100000 + 1/\ 2$	$3/3$

many real players will choose 0. Let us consider Alice at the state parametrized by $(\alpha_0, \alpha_1) \approx (0, 0)$: here she feels that Bob will choose 1_B with probability nearly 1. (Alice may be a believer in the classical game theory.) When the game I is given, Alice will choose the rational "1_A" with high probability, because the boundary of Fig. 6.6 is

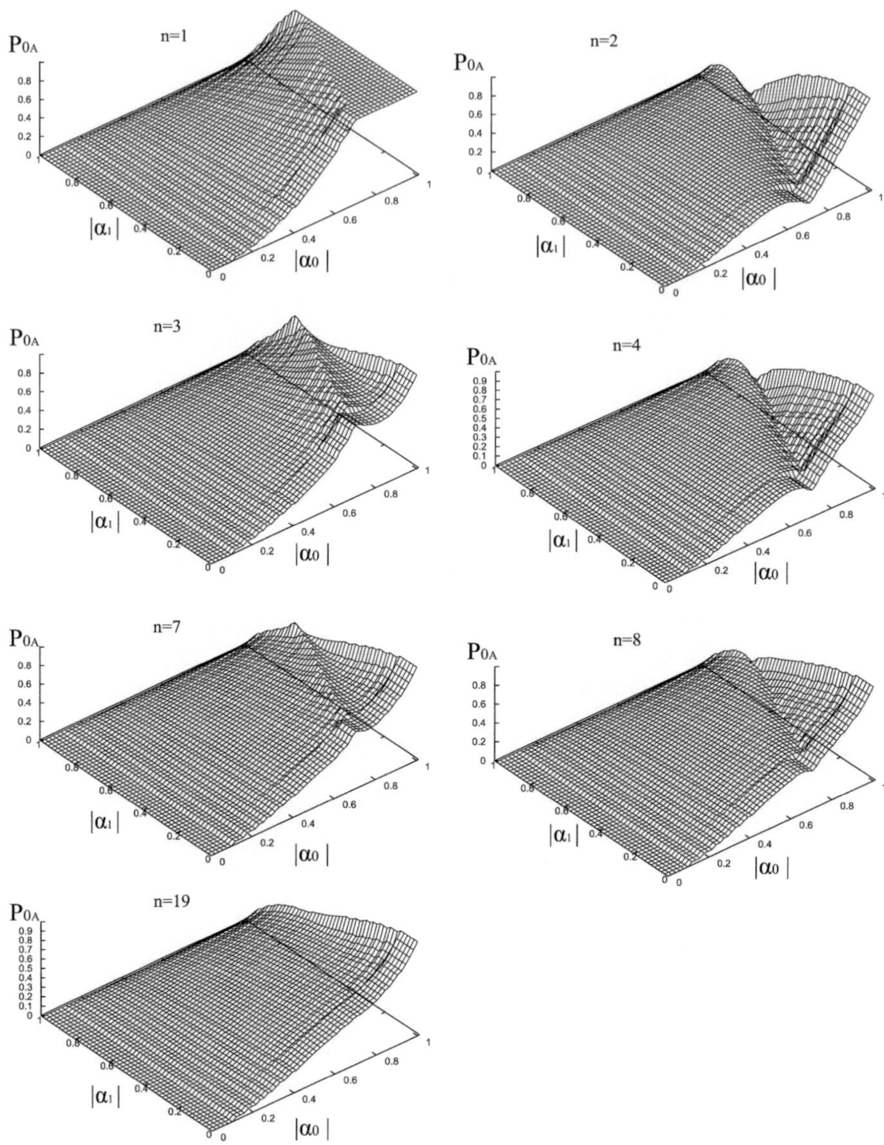

Fig. 6.7 The probability $P_{0_A}(n)$ fluctuated in the process of decision making

far from Alice's point $(\alpha_0, \alpha_1) \approx (0, 0)$. In the game III or the above extreme game, Alice may touch the boundary and develop the possibility of the irrational choice.

Fig. 6.8 The boundaries in PD game II and III

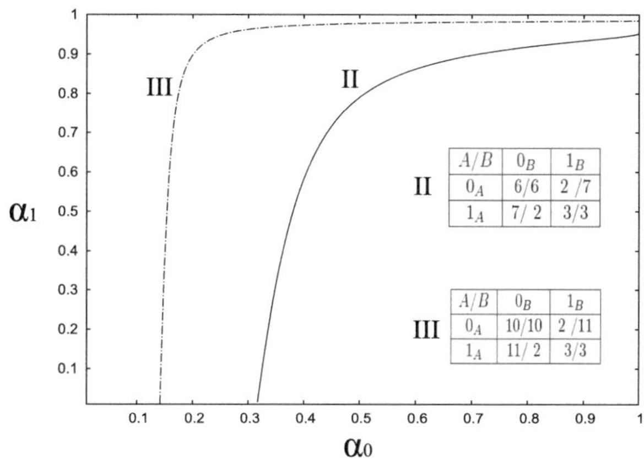

6.3 Bayesian Updating Biased by Psychological Factors

In this section, we introduce an example of the use of non-Kolmogorov probabilities in inference problems.

Let us think of the following story. There was a car accident in a small town, but the hitting car ran away from the scene of the accident. In the town there are only three yellow cars and seven blue cars. Can you estimate the probability (or possibility) of the color of the suspected car? Some people say that the probability of a yellow car is 0.3 for the statistical data in the town. However it is not necessary to use this statistical data since we do not have enough information to justify its use. Actually, other people may think that for some reason, yellow cars are more dangerous than blue ones. For example, if we know another car accident caused by a blue car just several days before, then we feel that a blue car is suspicious. This kind of probability is not obtained from statistically repeated trials, but from subjective inference of humans. Therefore, in this sense, it is called *subjective probability*.

Here let us think of the following case: a witness appears and he or she says "The hitting car was blue!!". If we believe in the witness' evidence, then we can conclude that the suspected car was blue. However, if we think that the witness may have misunderstood the color with some probability (error rate), our inference will be still probabilistic. Then another problem arises: how large is the value of the probability after we have known the information from the witness? According to the classical probability theory, this posterior probability of blue or yellow must be calculated by Bayes' formulae.[2] This method with Bayes' formulae is called *Bayesian updating*. Bayesian updating is one of important research areas in statistics, which is also widely used in computer engineering, e.g., machine learning.

[2] Of course, every probability in this case is subjective, so that we do not need to use classical probability theory in order to estimate this posterior probability. Instead of Bayes' formulae, humans often use more heuristic inference, and there are many reports in psychology.

Bayesian updating is used to describe a process of inference, which is explained as follows. Consider two event systems denoted by $S_1 = \{A, B\}$ and $S_2 = \{C, D\}$, where the events A and B are mutually exclusive, and the same holds for C and D. First, a decision making entity, say Alice, estimates the probabilities $P(A)$ and $P(B)$ for the events A and B, which are called the *prior probabilities*. The prior probability is sometimes called "*subjective* probability" or "*personal* probability". Further, Alice knows the conditional probabilities $P(C|A)$ and $P(C|B)$ obtained from some statistical data. When Alice sees the occurrence of the event C or D in the system S_2, she can change her prior prediction $P(A)$ and $P(B)$ to the following conditional probabilities by *Bayes' rule*: When Alice sees the occurrence of C in S_2, she can update her prediction for the event A from $P(A)$ to

$$P(A|C) = \frac{P(C|A)P(A)}{P(C|A)P(A) + P(C|B)P(B)}.$$

When Alice sees the occurrence of D in S_2, she can update her prediction for the event A from $P(A)$ to

$$P(A|D) = \frac{P(D|A)P(A)}{P(D|A)P(A) + P(D|B)P(B)}.$$

In the same way she updates her prediction for the event B. These conditional (updating) probabilities are called the *posterior probabilities*. The change of prediction is described as "updating" from the prior probabilities $P(A)$, $P(B)$ to the posterior probabilities, and it is a classical scheme of Bayesian updating.

In the paper [43], we redescribed the process of Bayesian updating in the framework of "quantum-like representation", where we introduced the following state vector belonging to the Hilbert space $\mathcal{H} = \mathcal{H}_1 \otimes \mathcal{H}_2 = \mathbb{C}^2 \otimes \mathbb{C}^2$;

$$\begin{aligned} |\Phi\rangle &= \sqrt{P(A')}\,|A'\rangle \otimes (\sqrt{P(C'|A')}\,|C'\rangle + \sqrt{P(D'|A')}\,|D'\rangle) \\ &\quad + \sqrt{P(B')}\,|B'\rangle \otimes (\sqrt{P(C'|B')}\,|C'\rangle + \sqrt{P(D'|B')}\,|D'\rangle). \end{aligned} \tag{6.23}$$

We call this vector the *prediction state vector*. The set of vectors $\{|A'\rangle, |B'\rangle\}$ is an orthogonal basis on $\mathcal{H}_1$, and $\{|C'\rangle, |D'\rangle\}$ is another orthogonal basis on $\mathcal{H}_2$. The A', B', C' and D' represent the events defined as

Event A' (B'): Alice **judges** "*the event* $A(B)$ *occurs in the system* S_1."
Event $C'(D')$: Alice **judges** "*the event* $C(D)$ *occurs in the system* S_2."

These events are *subjective* events (judgments) in *Alice's "mental space"* and the vectors $|A'\rangle$, $|B'\rangle$, $|C'\rangle$ and $|D'\rangle$ give a quantum-like representation of the above judgments. The vector $|\Phi\rangle$ represents the coexistence of these judgments in Alice's brain. For example, Alice is conscious of $|A'\rangle$ with the weight $\sqrt{P(A')}$, and under the condition of the event A', she sets the weights $\sqrt{P(C'|A')}$ and $\sqrt{P(D'|A')}$ for the judgments $|C'\rangle$ and $|D'\rangle$. Such an assignment of weights implies that Alice feels *causality* between S_1 and S_2: The events in S_1 are *causes* and the events in S_2 are *results*. The square of $\sqrt{P(A')}$ corresponds to a prior probability $P(A)$ in the Bayesian theory. If Alice knows the objective conditional probabilities $P(C|A)$ and $P(C|B)$, she can set the weights of $\sqrt{P(C'|A')}$ and $\sqrt{P(C'|B')}$ from $P(C'|A') = P(C|A)$ and $P(C'|B') = P(C|B)$. If Alice has the prediction state $|\Phi\rangle\langle\Phi| \equiv \rho$ and sees the occurrence of the event C in S_2, the event D' *vanishes* instantaneously (in her mental representation). This vanishing is represented as the reduction by the projection operator $M_{C'} = I \otimes |C'\rangle\langle C'|$;

$$\frac{M_{C'}\rho M_{C'}}{\mathrm{tr}(M_{C'}\rho)} \equiv \rho_{C'}$$

The posterior probability $P(A|C)$ is calculated by

$$\mathrm{tr}(M_{A'}\rho_{C'}),$$

where $M_{A'} = |A'\rangle\langle A'| \otimes I$.

The inference based on the Bayesian updating is rational—from the viewpoint of classical probability theory, game theory and economics (the Savage sure thing principle). However, in cognitive psychology and economics one can find extended empirical data showing that sometimes human inference seems to be irrational, see [23] for the review. Typically this happens in contexts such that there are (often hidden) psychological factors disturbing the rational inferences. Our aim is to provide a mathematical description of such an irrational inference; the concept of lifting will be used. Let us introduce a lifting from $S(\mathscr{H})$ to $S(\mathscr{H} \otimes \mathscr{K})$ by

$$\mathscr{E}_{\sigma V}(\rho) = V\rho \otimes \sigma V^*.$$

Here $\sigma \in S(\mathscr{K})$ represents the state of Alice's psychological representation of context, which is generated when Alice updates her inference. The operator V on $\mathscr{H} \otimes \mathscr{K}$ is unitary and gives the correlation between the prediction state ρ and the psychological factor σ, in other words, it specifies psychological affection to the rational inference. We call the state defined by

$$\rho_{\sigma V} \equiv \mathrm{tr}_{\mathscr{K}}(\mathscr{L}_{\sigma V}(\rho)),$$

the prediction state biased from the rational prediction ρ. From this $\rho_{\sigma V}$, the joint probability is defined as

$$\mathrm{tr}\,(M_{X'}M_{Y'}\rho_{\sigma V}M_{Y'}) \equiv P_{\sigma V}(X', Y'),$$

for the events $X' = (A'$ or $B')$ and $Y = (C'$ or $D')$, and *the biased posterior probability* is defined as

$$P_{\sigma V}(X'|Y') = \frac{P_{\sigma V}(X', Y')}{\mathrm{tr}(M_{C'}\rho_{\sigma V})}$$

In general, the value of $P_{\sigma V}(X'|Y')$ is different from the original $P(X'|Y')$ obtained with the aid of the rational (Bayesian) inference.

6.4 Optical Illusions for an Ambiguous Figure

In this section, we discuss the process of visual perception of human beings from the adaptive dynamical point of view.

6.4.1 What Are Optical Illusions?: A Mystery of Visual Perception

Most people may have the experience that when they watch the moon in the evening, the moon near the horizon is larger than when it is high in the sky. Nevertheless, in reality, the size of the moon does not change. This phenomenon has been known since the ancient Greek era, and nowadays it is called moon illusion. Why do we feel such an unreality? Many psychologists or other scientists have been discussing the mechanism of such an illusion, and several hypotheses have been proposed so far; for example, the moon illusion is caused by the atmospheric refraction, see the Ptolemy's book of *Almagest* (ca. 150 C.E.), or by the movement of eyes [44], or by others physiological factors. However, all such hypotheses are not definitive, and the debate is still in progress.

What we can clearly say is that our perception of an object is not exactly the same as it exists in reality. A human being does not recognize just a raw snapshot of what comes into our eyes. Our recognition is unconsciously biased. How to bias depends on how it exists in certain surroundings. In Fig. 6.9, there are two pictures which have a single letter "A". These letters are of the same brightness of color in reality. However they seem different to us, which is biased by the background colors. This is also one of the optical illusions.

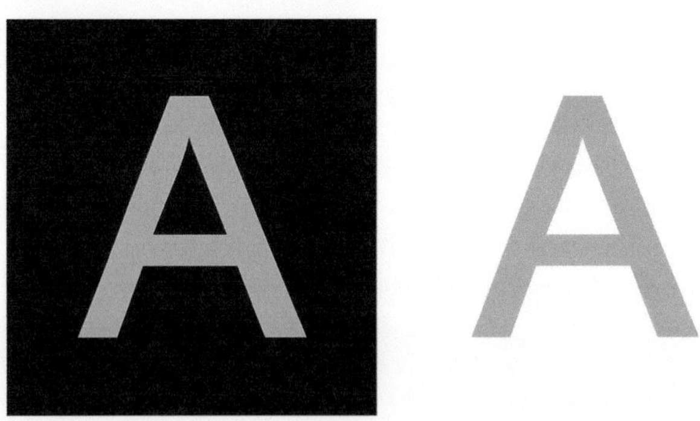

Fig. 6.9 Perceptual constancy: we can see the different brightness of the letter 'A' in the different contrast of background

Psychologists explain that this is caused by *perceptual constancy*. The perceptual constancy plays a very important role in human recognition. When we see a pyramid, the shape of pyramid is different depending on the angles at which we see it. From the side we can see a triangle, while from above we can see a square. However, we can recognize that this triangle and this square represent the same object (a pyramid). It is considered that such stable recognitions of humans are assisted by the perceptual constancy.

As explained above, optical illusions are essential phenomena in human recognition process. In the next section, we shall discuss an example of optical illusions in the framework of the theory of adaptive dynamics which was presented in the previous chapters.

6.4.2 Optical Illusion for Ambiguous Figure

An ambiguous figure is a figure which has multiple different meanings. See Fig. 6.10, called *Schröder's stair*. The ambiguous figure induces optical illusion [45]. Our brain can switch between the two alternative interpretations of this figure; (i) The surface of 'L' is at the front, and the surface of 'R' is at the back; (ii) The surface of 'R' is at the front, and the surface of 'L' is at the back. This switch-like process of human perception is called *depth inversion*, and many experimental proofs of this phenomenon have been reported. However, similarly to other optical illusions, the details of its mechanism are not completely figured out even in recent studies.

Recently, Atmanspacher et al. have explained the frequent change of human perception for another ambiguous figure, which is called Necker's cube by means of quantum-like model on two-level system [46]. They introduce the neutral state for Necker's cube, that is, they consider the pre-observed state, which is not changed to one of two perceptions. The mathematical expression of the neutral state is given by a superposition of two orthogonal vectors in the two dimensional space, and its oscillation is given by a certain Hamiltonian, a Pauli matrix. They also consider frequent measurements given by two projections with respect to the orthogonal vectors. By a series of measurements, they described a stable recognition of the figure.

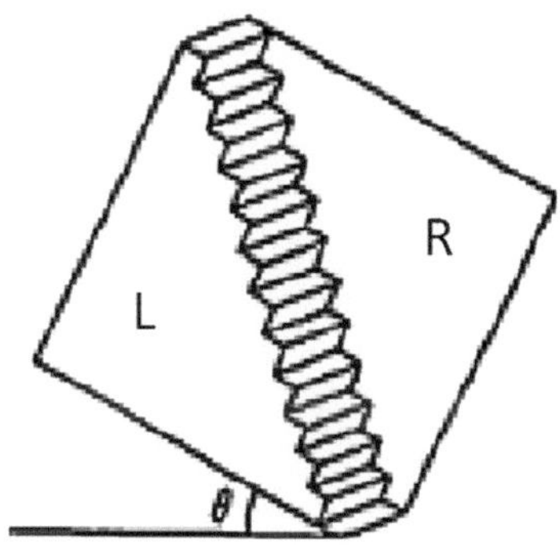

Fig. 6.10 Schröder's stair leaning at angle θ

On the other hand, we focus on the context of the perception process from the adaptive-dynamical point of view. Let us come back to the discussion on Schröder's stair. It is a well-known fact that the depth inversion depends on various contexts of the figure; e.g., the relative size of the surface 'L' for 'R', color or shadow of the figure, the angle to the horizon, etc. We discuss the adaptive system where such contextual dependencies emerge. We calculate the contextual dependent probability that a person answers either (i) or (ii) in the experiment.

We showed the picture of Schröder's stair which is inclined at a certain angle θ (see Fig. 6.10) to the 151 subjects. We prepared the 11 pictures which are inclined at different angles: $\theta = 0, 10, 20, 30, 40, 45, 50, 60, 70, 80, 90$. A subject must answer either (i) "L is the front side" or (ii) "R is the front side" for every picture. We arranged the computer experiment to change the pictures and to record their answers.

Before the experiment, we divided the 151 subjects into three groups: (A) 55 persons, (B) 48 persons, (C) 48 persons. For the first group (A), the order of showing is randomly selected for each person. To assume statistically uniform randomness of this selection, we used the computer-implemented function (e.g. java.rand). For the second group (B), the angle θ is changed in the increasing sequence: $0, 10, \ldots, 90$. Inversely, for the third group (C), the angle θ is changed in the decreasing sequence: $90, 80, \ldots, 0$.

Figure 6.11 shows the probability that a subject says (i) "L is the front side" with respect to angle θ. We denote it by $P(X_\theta = L)$. The probability $P(X_\theta = L)$ is decreased as the value of θ is increased. However, the speed of decreasing is different in the three cases. Moreover, we can see that the probability at $\theta = 0$ is nearly equal to one in each cases. That is, almost every subject says "L is the front side" at the angle $\theta = 0$. Conversely, the probability $\theta = 90$ is nearly equal to zero, that is, almost every subject says "R is the front side" at the angle $\theta = 90$. Therefore it can be considered that subjects feel little ambiguity of the picture for $\theta = 0$ and 90. In fact, we observed that it took very short time (less than 1 s) to make his/her decision when the angle θ is around 0 or 90°.

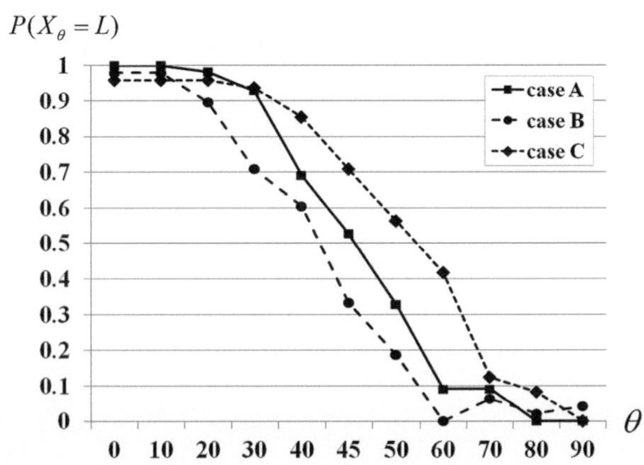

Fig. 6.11 Probability $P(X_\theta = L)$ with respect to θ

6.4.3 Violation of Formula of Total Probability

In fact, the experiments on recognition of ambiguous figures were the first experiments outside of physics (in cognitive psychology), which were designed to demonstrate violation of the formula of total probability, see [8, 9]. In this section we show that the data from the experiment with the Schröder stair, Sect. 6.4.2, also violate this basic law of the Kolmogorov theory of probability.

As explained in the previous section, we consider the three situations (A), (B) and (C) in the experiments. We can write the violation of total probability law as follows:

$$P_A(X_\theta = L) = P_B(X_\theta = L, X_{\theta'} = L) + P_B(X_\theta = L, X_{\theta'} = R) + \Delta_{AB},$$
$$P_A(X_\theta = L) = P_C(X_\theta = L, X_{\theta'} = L) + P_C(X_\theta = L, X_{\theta'} = R) + \Delta_{AC},$$
$$P_C(X_\theta = L) = P_B(X_\theta = L, X_{\theta'} = L) + P_B(X_\theta = L, X_{\theta'} = R) + \Delta_{CB},$$

where θ and θ' are different angles. Note that, with experimental results, we can estimate the degree of violation Δ_{AB}, Δ_{AC} and Δ_{CB}, for instance, Δ_{AB} at the angle θ is given by

$$\Delta_{AB} = P_A(X_\theta = L) - P_B(X_\theta = L, X_{\theta'} = L) + P_B(X_\theta = L, X_{\theta'} = L)$$
$$= P_A(X_\theta = L) - P_B(X_\theta = L).$$

We can see that Δ_{AB} does not depend on θ' since only θ appears in the RHS of the above equation. This comes from Kolmogorovness within the context B s.t.

$$P_B(X_\theta = L) = P_B(X_\theta = L, X_{\theta'} = L) + P_B(X_\theta = L, X_{\theta'} = L).$$

From similar discussions, one can find that Δ_{AC} and Δ_{CB} depend only on θ but not on θ'. As seen in Fig. 6.12, the strong violation becomes manifest in the middle range of the angle θ.

Here, let us compare the above Δ_{AB}, Δ_{AC} or Δ_{BC} with the conventional trigonometric interference in quantum mechanics. If Δ_{AB}, Δ_{AC} and Δ_{BC} have the form of

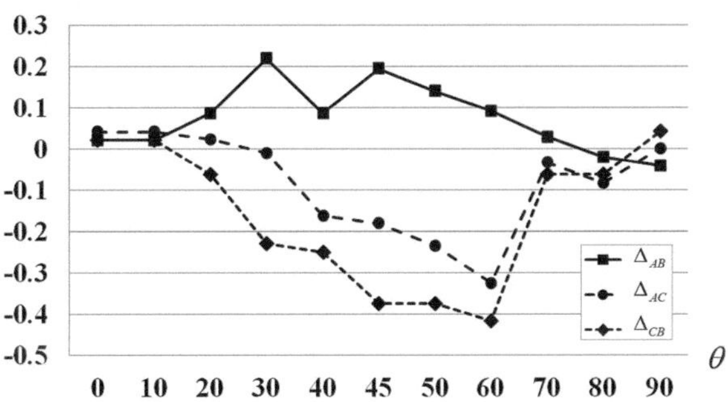

Fig. 6.12 Degree of violation of total probability law

trigonometric interference

$$\begin{aligned}
\Delta_{AB} &= 2\cos(\phi_{AB})\sqrt{P_B(X_\theta = L, X_{\theta'} = L)P_B(X_\theta = L, X_{\theta'} = R)},\\
\Delta_{AC} &= 2\cos(\phi_{AC})\sqrt{P_C(X_\theta = L, X_{\theta'} = L)P_C(X_\theta = L, X_{\theta'} = R)},\\
\Delta_{BC} &= 2\cos(\phi_{BC})\sqrt{P_C(X_\theta = L, X_{\theta'} = L)P_C(X_\theta = L, X_{\theta'} = R)}, \qquad (6.24)
\end{aligned}$$

then the following quantities should take values between -1 and $+1$.

$$\begin{aligned}
\delta_{AB} &= \frac{\Delta_{AB}}{2\sqrt{P_B(X_\theta = L, X_{\theta'} = L)P_B(X_\theta = L, X_{\theta'} = R)}},\\
\delta_{AC} &= \frac{\Delta_{AC}}{2\sqrt{P_C(X_\theta = L, X_{\theta'} = L)P_C(X_\theta = L, X_{\theta'} = R)}},\\
\delta_{BC} &= \frac{\Delta_{BC}}{2\sqrt{P_C(X_\theta = L, X_{\theta'} = L)P_C(X_\theta = L, X_{\theta'} = R)}}
\end{aligned}$$

However, as shown in Fig. 6.13, the values of δ_{BC} exceed one at several points (θ, θ'). Therefore, the probability $P_B(X_\theta = L)$ can not be written as the conventional form of quantum mechanical interference:

$$P_B(X_\theta = L) \neq \Big|\sqrt{P_C(X_\theta = L, X_{\theta'} = L)} + \exp(i\phi_{BC})\sqrt{P_C(X_\theta = L, X_{\theta'} = R)}\Big|^2.$$

In order to explain δ_{BC} exceeding 1, we can use the generalized description of quantum mechanical interference, which has been discussed in [23].

These generalized Born's rules can be applied to our experimental result. However, the interpretation of such generalization is not figured out well. In the next section, we give more specific model based on the psychological process of self-dialog.

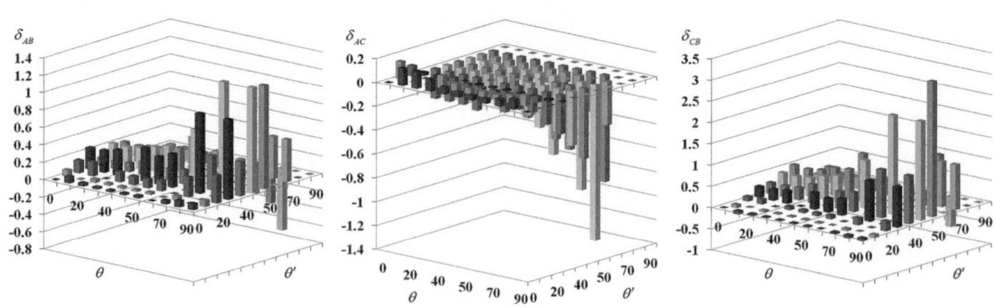

Fig. 6.13 Violation of total probability law

6.4.4 A Model of Depth Inversion: Majority Among N-agent

Subjects have to answer either (i) "L is the front side" or (ii) "R is the front side", so that we take the Hilbert space $\mathcal{H} = \mathbb{C}^2$ for this model. Let $|L\rangle = \begin{pmatrix} 1 \\ 0 \end{pmatrix}$ and $|R\rangle = \begin{pmatrix} 0 \\ 1 \end{pmatrix}$ be the orthogonal vectors describing the answers (i) and (ii), respectively. The initial state of a subject's mind is given by

$$\rho \equiv |x\rangle \langle x|,$$

where $|x\rangle$ is a state vector $\frac{1}{\sqrt{2}}(|L\rangle + |R\rangle)$, which represents the neutral mind for $|L\rangle$ and $|R\rangle$ before the decision making starts.

When a subject is shown a picture inclined at an angle θ, he recognizes the lean of the picture. Such a recognition process is given by the operator

$$M(\theta) = \begin{pmatrix} \cos\theta & 0 \\ 0 & \sin\theta \end{pmatrix}.$$

After the recognition, the state of mind changes from the initial state ρ to an adaptive state

$$\rho_\theta = \Lambda_M^*(\rho) \equiv \frac{M^*\rho M}{\mathrm{tr}\left(|M|^2 \rho\right)} = \begin{pmatrix} \cos^2\theta & \cos\theta\sin\theta \\ \cos\theta\sin\theta & \sin^2\theta \end{pmatrix}.$$

The fluctuation between (i) and (ii) is expressed as ρ_θ above.

Here, we introduced the following assumption of a self-dialog process by a subject. At the beginning of the process, the subject imagines and creates an imaginary agent in the brain, and this agent has its own mental state which is expressed as the adaptive state ρ_θ. The subject repeatedly creates imaginary agents during the experiment for X_θ. We can consider the adaptive state describing the "agents" involved in the process of self-dialog as the N-composite state of ρ_θ:

$$\sigma = \underbrace{\rho_\theta \otimes \cdots \otimes \rho_\theta}_{N},$$

where N is the number of the agents created by the subject in the process of approaching to the answer (i) or (ii). If it takes him little time to answer, then N might be small. After the creation of agents (or on the way to create them), the subject can talk with the imaginary agents in the brain, and he/she knows the answer of each agent. Through this dialogue, the subject can know the opinions of all the agents. For N agents, there are 2^N possible opinions. The subject's answer is determined by reference to the opinions of all the agents.

At the final step of this dialogue, we additionally assumed that σ changed into $\left|L^{\otimes N}\right\rangle$ or $\left|R^{\otimes N}\right\rangle$ since no subject in this experiment can answer anything except (i) or (ii). Under this assumption, we introduce the observable-adaptive operator Q, which describes the process of making a common decision. For example, in the case of $N = 2$, 3 and 4, the operator Q is

$$Q^{(2)} = |LL\rangle\langle LL| + |RR\rangle\langle RR|,$$

$$\begin{aligned} Q^{(3)} &= |LLL\rangle\left(\langle LLL| + \langle LLR| + \langle LRL| + \langle RLL|\right) \\ &+ |RRR\rangle\left(\langle RRR| + \langle RRL| + \langle RLR| + \langle LRR|\right), \end{aligned}$$

$$\begin{aligned} Q^{(4)} &= |LLLL\rangle\left(\langle LLLL| + \langle LLLR| + \langle LLRL| + \langle LRLL| + \langle RLLL|\right) \\ &+ |RRRR\rangle\left(\langle RRRR| + \langle RRRL| + \langle RRLR| + \langle RLRR| + \langle LRRR|\right). \end{aligned}$$

By applying the operator Q to the state of agents σ, the minority of opinions of the agents are ignored and changed to the opinion of the majority agents. In this sense, our model is considered as the *majority voting system* for the N agents.

Here, the lifting $\mathcal{E}^*_{\sigma Q} : \mathcal{S}\left(\mathbb{C}^2\right) \to \mathcal{S}\left(\mathbb{C}^{2N}\right)$ is defined as

$$\mathcal{E}^*_{\sigma Q}(\rho) = \frac{Q\sigma Q^*}{\operatorname{tr}\left(|Q|^2\sigma\right)} = \frac{Q\left\{\Lambda^*_M(\rho)\right\}^{\otimes N} Q^*}{\operatorname{tr}\left(|Q|^2\left\{\Lambda^*_M(\rho)\right\}^{\otimes N}\right)} \in \mathcal{S}\left(\mathbb{C}^{2N}\right),$$

and the probabilities for X_θ are given with this lifting as

$$\begin{aligned} P(X_\theta = L) &\equiv \operatorname{tr}\left(|L\rangle\langle L|^{\otimes N}\,\mathcal{E}^*_{\sigma Q}(\rho)\right), \\ P(X_\theta = R) &\equiv \operatorname{tr}\left(|R\rangle\langle R|^{\otimes N}\,\mathcal{E}^*_{\sigma Q}(\rho)\right). \end{aligned}$$

In $N = 2$, 3 and 4, the probability $P(X_\theta = L)$ has the following forms:

$$P^{(2)}(X_\theta = L) = \frac{\cos^4\theta}{\cos^4\theta + \sin^4\theta},$$

$$P^{(3)}(X_\theta = L) = \frac{\left(\cos^3\theta + 3\cos^2\sin\theta\right)^2}{\left(\cos^3\theta + 3\cos^2\sin\theta\right)^2 + \left(\sin^3\theta + 3\sin^2\cos\theta\right)^2},$$

$$P^{(4)}(X_\theta = L) = \frac{\left(\cos^4\theta + 4\cos^3\theta\sin\theta\right)^2}{\left(\cos^4\theta + 4\cos^3\theta\sin\theta\right)^2 + \left(\sin^4\theta + 4\sin^3\cos\theta\right)^2}.$$

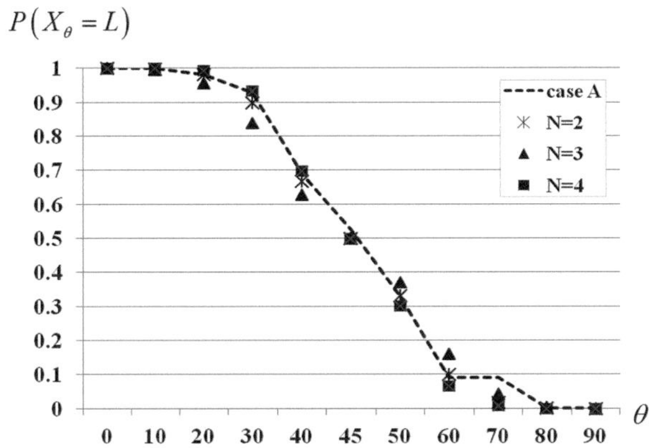

Fig. 6.14 Values of $P\,(X_\theta = L)$ in our model and its comparison with experimental data (A)

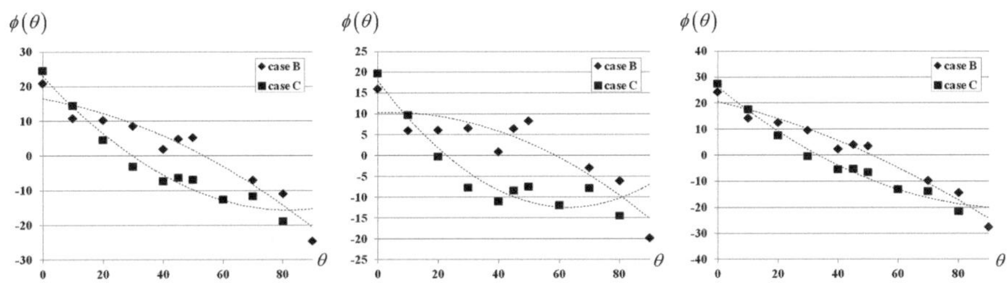

Fig. 6.15 Noise of biased phase; (*Left*) $N = 2$, (*Center*) $N = 3$, (*right*) $N = 4$

We compared these probabilities with the experimental data of case (A), see Fig. 6.14. One can find that the probabilities $P(X_\theta = L)$ in $N = 2$, 3 and 4 coincide with the experimental data of A.

In the situation (B) or (C), we take another adaptive operator $M(\theta + \phi)$, whose angle is shifted by an unknown psychological bias ϕ instead of $M(\theta)$. This value of ϕ can also be calculated by the experimental data. We show this in Fig. 6.15. The values of ϕ in (B) are higher than those in (C), and especially the difference between $\phi^{(B)}$ and $\phi^{(C)}$ is clearly seen in the middle range of the angle θ. By increasing N, this difference becomes smaller.

6.5 Contextual Leggett-Garg Inequality for Statistical Data on Recognition of Ambiguous Figures

In physics the Leggett-Garg (LG) inequality [47] is one of the fundamental statistical tests demonstrating the impossibility to use the classical realistic model in quantum physics. This is a kind of the Bell inequality [48]. Opposite to spatial versions of the Bell inequality emphasizing the role of nonlocality (such as Clauser-Horne-

Shimony-Holt inequality [49]), the LG-inequality directly questions the possibility of the realistic description. Unfortunately, the problem of realism is shadowed by the assumption of non-invasiveness of measurements, condition (A2) below. The latter is a complicated issue even in physics; in biology the situation is even more complicated. Therefore one cannot apply directly the standard physical interpretation of violation of the LG-inequality to biology. One has to find a new interpretation matching biological applications.

We interpret the Leggett-Garg (LG) inequality as a kind of *contextual probabilistic inequality*, in which one combines the data collected in experiments performed for three different contexts.[3] In the original version of the inequality [47] these contexts have the temporal nature and they are given by three pairs of instances of time, $(t_1, t_2), (t_2, t_3), (t_3, t_4)$, where $t_1 < t_2 < t_3$. We reinterpret the LG conditions of *macroscopic realism* and *noninvasive measurability* in the general contextual framework. Our formulation is done in the purely probabilistic terms: the existence of the context independent joint probability distribution P and the possibility to reconstruct the experimentally found marginal (two dimensional) probability distributions from P. We derive an analog of the LG inequality, "contextual LG inequality", and use it as a test of "quantum-likeliness" of the statistical data collected in a series of experiments on recognition of ambiguous figures, see also Sect. 6.4.3 on violation of FTP for recognition of ambiguous figures. In our experimental study on the LG-inequality the figure under recognition is the Schröder stair, which is shown with rotations for different angles, Sect. 6.4.2. The contexts are encoded by dynamics of rotations: clockwise, anticlockwise, and random, see again Sect. 6.4.2. Our data demonstrated violation of the contextual LG inequality for some combinations of the aforementioned contexts. Since in quantum theory and experiments with quantum physical systems this inequality is violated, e.g., in the form of the original LG-inequality, our result can be interpreted as a sign that the quantum(-like) models can provide a more adequate description of the data generated in the process of recognition of ambiguous figures.

We also mention the quantum-like study of recognition of ambiguous figures done by Atmanspacher et al. [46, 52]: a QL model of bistable perception. (A generalized quantum formalism was in use [53].) These papers were pioneer works on applications of the LG-inequality [47] to cognitive science and decision making.

As has been emphasized throughout this book, in probabilistic terms the context dependence of (probabilistically determined) mental states implies that the conventional model of probability theory, the Kolmogorov measure-theoretic model [54], cannot be applied to describe statistics of recognition of ambiguous figures, see also Sect. 6.4.3. Thus our result on violation of the contextual LG inequality restricts the domain of applications of the Kolmogorov model. More general probabilistic models have to be tested, e.g., quantum probability and its generalizations, cf. also [53, 55].

[3] Similar interpretation for the Bell inequality was presented in the paper [50], the first paper which demonstrated that the Bell-type inequalities can be violated outside of physics, for the data collected from recognition of ambiguous figures; see also [51]; see [49] for the detailed presentation of the contextual viewpoint on quantum theory in general and on the Bell-type inequalities in particular.

6.5.1 *Leggett-Garg Inequality*

Conditions of Derivation

At the beginning of the discussion in the paper [47], Leggett and Garg (LG) postulated the following two assumptions:

(A1) **Macroscopic realism**: *A macroscopic system with two or more macroscopically distinct states available to it will at all times be in one or the other of these states.*

(A2) **Noninvasive measurability**: *It is possible, in principle, to determine the state of the system without arbitrarily small perturbation on its subsequent dynamics.*

Under these assumptions, the correlation functions must satisfy the LG inequality, which will be presented in the next section. However, quantum mechanics violates the LG inequality as well as spatial versions of the Bell inequality such as Clauser-Horne-Shimony-Holt inequality. Therefore this violation means that at least one of the two assumptions fails for quantum systems.

Although in the derivation of the LG inequality, Sect. 6.5.1, both conditions play important roles, their foundational values are different. The main issue is realism, whether one can still proceed with (A1), macroscopic realism, in the quantum world. Therefore the main part of the LG paper [47] is devoted to the discussion about possible physical experimental schemes, which may lead to noninvasive measurements or at least measurements in which invasiveness is small as compared with the degree of violation of the LG inequality. There are claims, e.g., the experiment in [56], that such negligibly invasive measurements were performed experimentally and the LG inequality was violated. This is an important argument in favor of non-objectivity of quantum observables.

However, the LG approach plays an important foundational role even if the possibility that measurements are non-negligibly invasive cannot be excluded. We know that classical systems and measurements on such systems satisfy conditions (A1) and (A2); e.g., airplane's trajectory. Therefore by violating LG we at least know that a phenomenon under study cannot be described classically.

In cognitive science it is not easy (if possible at all) to come with an experimental scheme, which would lead to (at least approximately) noninvasive measurements. The brain is a kind of self-measurement device, by giving an answer to a question the brain definitely perturbs its mental state non-negligibly. And the introspective measurements definitely have the lowest degree of non-invasiveness. Therefore it seems that violations of the LG type inequalities for the data collected, e.g., in cognitive psychology, cannot lead to the conclusion that mental realism is questionable. Here realism is understood in the sense of objectivity of mental observables: their values can be assigned, say to the brain, a priori, i.e., before measurements. Nevertheless, such violations show that the data under consideration is nonclassical, i.e., it is not similar to the data collected, e.g., from an ensemble of moving airplanes.

However, our main point is that in relation to the problem of cognition the standard physical viewpoint on conditions for derivation of the LG inequality, namely, the mixture of macrorealism and non-invasiveness, does not match the mental situation well enough. As was emphasized, the very notion of (non)invasive measurement loses its clearness for self-measuring devices, and the brain is one of such devices. This book advertises the contextual viewpoint on mental phenomena. It seems that Bell type inequalities, including the temporal ones, can be used to distinguish between contextual and non-contextual realism and more generally (since mental processes are fundamentally random) contextual and non-contextual probabilistic representations. As will be seen from coming presentation, non-contextuality of the representation of probabilistic data implies constraints on the data in the form of various inequalities (or equalities such as FTP). By using the contextual representation a system (including the brain) can violate such constraints.

Contextual Viewpoint on the Proof of LG Inequality

To provide a possibility to compare the original LG inequality with our contextual generalization, see Sect. 6.5.2, and at the same time to add the contextual flavor to the LG approach, we present the original LG derivation by considering time as a context parameter.

Let Q be an observable quantity which takes the values $+1$ or -1. In the original discussion by LG, Q is the observable of position of a particle in the two potential wells. However we can discuss another two-level system, e.g., the spin-$\frac{1}{2}$ system.

The measurement of a two-level system is performed on a single system at different times $t_1 < t_2 < t_3$. We denote the observable at time t_k by Q_k $(k = 1, 2, 3)$. By repeating a series of three measurements, we can estimate the values of the correlation functions by

$$C_{ij} = \frac{1}{N} \sum_{n=1}^{N} q_i^{(n)} q_j^{(n)},$$

where $q_i^{(n)}$ (or $q_j^{(n)}$) is the result of the n-th measurement of Q_i (or Q_j). Note that the correlation between Q_i and Q_j takes the maximum value $C_{ij} = 1$ when $q_i^{(n)} q_j^{(n)}$ equals 1 for all the repeated trials. Here, consider the assumption A1, then the state of the system is determined at all times even when the measurement is not performed on the system. Therefore, the values of joint probabilities of Q_1, Q_2 and Q_3 are determined *a priori* at the initial time t_0. We denote it by the symbol $P_{i,j}(Q_1, Q_2, Q_3)$. Remark that the pairs of indexes i, j encode the situation that only two observables Q_i and Q_j are measured. In other words, the joint probability depends on the situations for which a pair of observables is measured. (We can consider pairs of indexes, instances of time, as parameters encoding three temporal contexts, $\mathcal{C}_{t_1 t_2}, \mathcal{C}_{t_1 t_3}, \mathcal{C}_{t_2 t_3}$, cf. Sect. 6.5.2.) However, if one considers A2, then the joint probabilities do not depend on temporal contexts:

$$P_{i,j}(Q_1, Q_2, Q_3) = P(Q_1, Q_2, Q_3) \quad \forall i, j$$

Then we have the following formula:

$$\begin{aligned}
P(Q_1, Q_2) &= \sum_{Q_3=\pm 1} P(Q_1, Q_2, Q_3), \\
P(Q_2, Q_3) &= \sum_{Q_1=\pm 1} P(Q_1, Q_2, Q_3), \\
P(Q_1, Q_3) &= \sum_{Q_2=\pm 1} P(Q_1, Q_2, Q_3).
\end{aligned}$$

Moreover, the correlation functions are written with the joint probabilities $P\left(Q_i, Q_j\right)$ as

$$\begin{aligned}
C_{ij} &= P\left(Q_i = 1, Q_j = 1\right) + P\left(Q_i = -1, Q_j = -1\right) \\
&\quad - P\left(Q_i = -1, Q_j = 1\right) - P\left(Q_i = 1, Q_j = -1\right) \\
&= 2\left\{P\left(Q_i = 1, Q_j = 1\right) + P\left(Q_i = -1, Q_j = -1\right)\right\} - 1.
\end{aligned}$$

We set $K = C_{12} + C_{23} - C_{13}$. It can be represented in the following form:

$$K = 1 - 4\{P(Q_1 = 1, Q_2 = -1, Q_3 = 1) + P(Q_1 = -1, Q_2 = 1, Q_3 = -1)\} \tag{6.25}$$

This representation implies the LG-inequality:

$$K \leq 1. \tag{6.26}$$

As we know, e.g., [47, 56] for quantum correlation functions C_{ij} the above inequality can be violated (theoretically and experimentally).

6.5.2 *Contextual Leggett-Garg Inequality*

Here, we express the LG's assumptions in terms of context-dependent probabilities. We remark that in general context-dependent probabilities cannot be represented in the common Kolmogorov probability space. Therefore one can consider such contextual probabilistic models as non-Komogorovian probabilistic models, see Appendix.

(A1) There exists joint probability $P_{\mathscr{C}}(Q_1, Q_2, Q_3)$ under a certain condition of experiments (context) $\mathscr{C}$. And the Kolmogorovness of $P_{\mathscr{C}}(Q_1, Q_2, Q_3)$ is ensured within the context $\mathscr{C}$:

$$P_{\mathscr{C}}(Q_1, Q_2) = \sum_{Q_3=\pm 1} P_{\mathscr{C}}(Q_1, Q_2, Q_3),$$

$$P_{\mathscr{C}}(Q_2, Q_3) = \sum_{Q_1=\pm1} P_{\mathscr{C}}(Q_1, Q_2, Q_3),$$

$$P_{\mathscr{C}}(Q_1, Q_3) = \sum_{Q_2=\pm1} P_{\mathscr{C}}(Q_1, Q_2, Q_3),$$

and

$$\begin{aligned}
P_{\mathscr{C}}(Q_1) &= \sum_{Q_2=\pm1} P_{\mathscr{C}}(Q_1, Q_2) = \sum_{Q_3=\pm1} P_{\mathscr{C}}(Q_1, Q_3) \\
&= \sum_{Q_2=\pm1} \sum_{Q_3=\pm1} P_{\mathscr{C}}(Q_1, Q_2, Q_3), \\
P_{\mathscr{C}}(Q_2) &= \sum_{Q_1=\pm1} P_{\mathscr{C}}(Q_1, Q_2) = \sum_{Q_3=\pm1} P_{\mathscr{C}}(Q_2, Q_3) \\
&= \sum_{Q_1=\pm1} \sum_{Q_3=\pm1} P_{\mathscr{C}}(Q_1, Q_2, Q_3), \\
P_{\mathscr{C}}(Q_3) &= \sum_{Q_2=\pm1} P_{\mathscr{C}}(Q_2, Q_3) = \sum_{Q_1=\pm1} P_{\mathscr{C}}(Q_1, Q_3) \\
&= \sum_{Q_1=\pm1} \sum_{Q_2=\pm1} P_{\mathscr{C}}(Q_1, Q_2, Q_3).
\end{aligned}$$

(A2) Consider three different contexts $\mathscr{C}_A$, $\mathscr{C}_B$ and $\mathscr{C}_C$, then there exists a context $\mathscr{C}$ unifying the above contexts $\mathscr{C}_A$, $\mathscr{C}_B$ and $\mathscr{C}_C$ such that

$$P_{\mathscr{C}_A}(Q_1, Q_2) = \sum_{Q_3=\pm1} P_{\mathscr{C}}(Q_1, Q_2, Q_3),$$

$$P_{\mathscr{C}_B}(Q_2, Q_3) = \sum_{Q_1=\pm1} P_{\mathscr{C}}(Q_1, Q_2, Q_3),$$

$$P_{\mathscr{C}_C}(Q_1, Q_3) = \sum_{Q_2=\pm1} P_{\mathscr{C}}(Q_1, Q_2, Q_3).$$

Under these assumptions, one can obtain inequality (6.25) for K given by

$$K = 1 - 4(P_{\mathscr{C}}(Q_1 = 1, Q_2 = -1, Q_3 = 1) + P_{\mathscr{C}}(Q_1 = -1, Q_2 = 1, Q_3 = -1)). \tag{6.27}$$

6.5.3 *Violation of Inequality in Optical Illusions*

In this section we use the data from the experiment presented in Sect. 6.4.2. Thus we have the three kinds of experimental data: (A) the angle of Schröder stair changes randomly, (B) from 0 to 90, (C) from 90 to 0. These contexts of experiments are denoted by $\mathscr{C}_A$, $\mathscr{C}_B$ and $\mathscr{C}_C$. Let X_θ be a random variable which takes the values ± 1. The event that a subject says "the left side is the front side" corresponds to the result $X_\theta = +1$. Then, from the repeated trials for each experimental context, we have the experimentally obtained values of joint probabilities:

$$P_{\mathscr{C}_A}(X_0, X_{10}, \ldots, X_{90}), \quad P_{\mathscr{C}_B}(X_0, X_{10}, \ldots, X_{90}), \quad P_{\mathscr{C}_C}(X_0, X_{10}, \ldots, X_{90}).$$

The correlation functions are given by

$$C_{12} = 2\left\{P_{\mathscr{X}}\left(X_{\theta_1} = 1, X_{\theta_2} = 1\right) + P_{\mathscr{X}}\left(X_{\theta_1} = -1, X_{\theta_2} = -1\right)\right\} - 1,$$
$$C_{23} = 2\left\{P_{\mathscr{Y}}\left(X_{\theta_2} = 1, X_{\theta_3} = 1\right) + P_{\mathscr{Y}}\left(X_{\theta_2} = -1, X_{\theta_3} = -1\right)\right\} - 1,$$
$$C_{13} = 2\left\{P_{\mathscr{Z}}\left(X_{\theta_1} = 1, X_{\theta_3} = 1\right) + P_{\mathscr{Z}}\left(X_{\theta_1} = -1, X_{\theta_3} = -1\right)\right\} - 1.$$

Here, the triple $(\mathscr{X}, \mathscr{Y}, \mathscr{Z})$ is given by a combination of the contexts $\mathscr{C}_A$, $\mathscr{C}_B$ and $\mathscr{C}_C$. We show the values of C_{12}, C_{23} and C_{13} in the Table 6.3.

We estimate the LHS of the inequality:

$$K(\theta_1, \theta_2, \theta_3) = C_{12} + C_{23} - C_{13}.$$

Tables 6.4 and 6.5 shows that the value of K with respect to $(\theta_1, \theta_2, \theta_3) = (0, 10, 20)$ and (40, 45, 50). The value of K exceeding one is seen in several cases.

The violation of the contextual LG inequality by the statistical data collected for observations of the Schröder stair rotated for different angles supports the *contextual cognition paradigm* presented in the previous chapters of this book. Biological probabilistic data are fundamentally contextual. The brain does not have a priori prepared "answers" to the question about the R/L structure of the Schröder stair for the fixed angle θ. Answers are generated depending on the mental context. Thus mental realism is a kind of contextual realism. There are practically no (at least not

Table 6.3 (Left) $(\theta_1, \theta_2, \theta_3) = (0, 10, 20)$, (Right) $(\theta_1, \theta_2, \theta_3) = (40, 45, 50)$

	C_{12}	C_{23}	C_{13}		C_{12}	C_{23}	C_{13}
$\mathscr{C}_A$	1.000	0.964	0.964	$\mathscr{C}_A$	0.091	0.091	0.127
$\mathscr{C}_B$	0.917	0.833	0.750	$\mathscr{C}_B$	0.375	0.625	0.083
$\mathscr{C}_C$	0.917	1.000	1.000	$\mathscr{C}_C$	0.625	0.375	0.167

Table 6.4 Value of K (0, 10, 20)

$\mathscr{X}, \mathscr{Y}, \mathscr{Z}$	$\mathscr{C}_A, \mathscr{C}_A$	$\mathscr{C}_A, \mathscr{C}_B$	$\mathscr{C}_A, \mathscr{C}_C$	$\mathscr{C}_B, \mathscr{C}_A$	$\mathscr{C}_B, \mathscr{C}_B$	$\mathscr{C}_B, \mathscr{C}_C$	$\mathscr{C}_C, \mathscr{C}_A$	$\mathscr{C}_C, \mathscr{C}_B$	$\mathscr{C}_C, \mathscr{C}_C$
$\mathscr{C}_A$	1.000	1.214	1.047	0.870	1.083	0.917	1.036	1.250	1.083
$\mathscr{C}_B$	0.917	1.130	0.964	0.786	1.000	0.833	0.953	1.167	1.000
$\mathscr{C}_C$	0.917	1.130	0.964	0.786	1.000	0.833	0.953	1.167	1.000

Table 6.5 Value of K (40, 45, 50)

$\mathscr{X}, \mathscr{Y}, \mathscr{Z}$	$\mathscr{C}_A, \mathscr{C}_A$	$\mathscr{C}_A, \mathscr{C}_B$	$\mathscr{C}_A, \mathscr{C}_C$	$\mathscr{C}_B, \mathscr{C}_A$	$\mathscr{C}_B, \mathscr{C}_B$	$\mathscr{C}_B, \mathscr{C}_C$	$\mathscr{C}_C, \mathscr{C}_A$	$\mathscr{C}_C, \mathscr{C}_B$	$\mathscr{C}_C, \mathscr{C}_C$
$\mathscr{C}_A$	1.000	1.214	1.047	0.870	1.083	0.917	1.036	1.250	1.083
$\mathscr{C}_B$	0.917	1.130	0.964	0.786	1.000	0.833	0.953	1.167	1.000
$\mathscr{C}_C$	0.917	1.130	0.964	0.786	1.000	0.833	0.953	1.167	1.000

so many), so to say, "absolute mental quantities", the "answers" to the same question vary essentially depending on context. This conclusion is not surprising in the framework of cognitive science and psychology, where various framing effects are well known. Thus the main contribution of this study is the demonstration of applicability of a statistical test of contextuality borrowed from quantum physics.

References

1. Fox, C., Rogers, B., Tversky, A.: Option traders exhibit subadditive decision weights. J. Risk Uncertain. **13**, 5–17 (1996)
2. Rottenstreich, Y., Tversky, A.: Unpacking, repacking and anchoring: advances in support theory. Psychol. Rev. **104**, 406–415 (1997)
3. Bearden, J.N., Wallsten, T., Fox, C.: Error and subadditivity: a stochastic model of subadditivity. University of Arizona—Departmentt Management and Policy, preprint (2005)
4. Tversky, A., Koehler, D.: Support theory: a nonextential representation of subjective probability. Psychol. Rev. **101**, 547–567 (1994)
5. Khrennikov, A., Haven, E.: The importance of probability interference in social science: rationale and experiment (2007). arXiv:0709.2802v1 [physics.gen-ph]
6. Khrennikova, P., Khrennikov, A., Haven, E.: The quantum-like description of the dynamics of party governance in the US political system (2012). arXiv:1206.2888 [physics.gen-ph]
7. Khrennikova, P.: Evolution of quantum-like modeling in decision making processes. AIP Conf. Proc. **1508**, 108–112 (2012)
8. Conte, E., Todarello, O., Federici, A., Vitiello, F., Lopane, M., Khrennikov, A.: A preliminary evidence of quantum-like behaviour in measurements of mental states. Quantum Theory: Reconsideration of Foundations. Series Mathematical Model in Physics, Engineering and Cognitive Science, vol. 10, pp. 679–702. Växjö University Press, Växjö (2004)
9. Conte, E., Khrennikov, A., Todarello, O., Federici, A., Zbilut, J.P.: Mental states follow quantum mechanics during perception and cognition of ambiguous figures. Open Syst. Inf. Dyn. **16**, 1–17 (2009)
10. Khrennikov, A.: Classical and quantum mechanics on information spaces with applications to cognitive, psychological, social and anomalous phenomena. Found. Phys. **29**, 1065–1098 (1999)
11. Khrennikov, A.: On quantum-like probabilistic structure of mental information. Open Syst. Inf. Dyn. **11**, 267–275 (2004)
12. Khrennikov, A.: Quantum-like brain: interference of minds. BioSystems **84**, 225–241 (2006)
13. Busemeyer, J.B., Wang, Z., Townsend, J.T.: Quantum dynamics of human decision making. J. Math. Psychol. **50**, 220–241 (2006)
14. Busemeyer, J., Bruza, P.D.: Quantum Cognition and Decision. Cambridge University Press, Cambridge (2012)
15. Hofstader, D. R.: Dilemmas for superrational thinkers, leading up to a luring lottery. Sci. Am. **6** (1983)

16. Hofstader, D.R.: Metamagical Themes: Questing for the Essence of Mind and Pattern. Basic Books, New York (1985)
17. Shafir, E., Tversky, A.: Thinking through uncertainty: nonconsequential reasoning and choice. Cogn. Psychol. **24**, 449–474 (1992)
18. Tversky, A., Shafir, E.: The disjunction effect in choice under uncertainty. Psychol. Sci. **3**, 305–309 (1992)
19. Croson, R.: The disjunction effect and reasoning-based choice in games. Organ. Behav. Hum. Dec. Process. **80**, 118–133 (1999)
20. Cheon, T., Takahashi, T.: Interference and inequality in quantum decision theory. Phys. Lett. A **375**, 100–104 (2010)
21. Cheon, T., Tsutsui, I.: Classical and quantum contents of solvable game theory on Hilbert space. Phys. Lett. A **348**, 147–152 (2006)
22. Khrennikov, A.: Quantum-like model of cognitive decision making and information processing. Biosystems **95**, 179–187 (2009)
23. Khrennikov, A.: Ubiquitous Quantum Structure: From Psychology to Finance. Springer, Heidelberg (2010)
24. Fichtner, K.-H., Fichtner, L., Freudenberg, W., Ohya, M.: On a quantum model of the recognition process. In: QP-PQ: Probability and White Noise Analysis, vol. 21, pp. 64–84 (2008)
25. Franco, F.: The conjunction fallacy and interference effects. J. Math. Psychol. **53**, 415–422 (2009)
26. Khrennikov, A., Haven, E.: Quantum mechanics and violations of the sure-thing principle: the use of probability interference and other concepts. J. Math. Psychol. **53**, 378–388 (2009)
27. Pothos, E.M., Busemeyer, J.R.: Can quantum probability provide a new direction for cognitive modeling? Behav. Brain Sci. **36**, 255–274 (2013)
28. Pothos, E. M., Busemeyer, J. R., Trueblood, J. S.: A quantum geometric model of similarity. Psychol. Rev. (in press)
29. Lambert-Mogiliansky, A., Busemeyer, J.R.: Emergence and instability of individual identity. In: Busemeyer, J.R., Dubois, F., Lambert-Mogiliansky, A., Melucci, M. (eds.) Quantum Interaction 6th International Symposium, QI 2012. Lecture Notes in Computer Science, vol. 7620, pp. 102–113 (2012)
30. Dzhafarov, E.N., Kujala, J.V.: Selectivity in probabilistic causality: where psychology runs into quantum physics. J. Math. Psychol. **56**, 54–63 (2012)
31. Dzhafarov, E.N., Kujala, J.V.: Random variables recorded under mutually exclusive conditions: contextuality-by-default, to appear in the Advances in Cognitive Neurodynamics. In: Proceedings of the 4th International Conference on Cognitive Neurodynamics (2013). arXiv:1309.0962 [quant-ph]
32. Wang, Z., Busemeyer, J.R.: A quantum question order model supported by empirical tests of an a priori and precise prediction. Top. Cogn. Sci. (2013) (to be published)
33. Wang, Zh, Busemeyer, J.R.: A quantum question order model supported by empirical tests of an a priori and precise prediction. Top. Cogn. Sci. **5**, 689–710 (2013)
34. D'Ariano, M., Fei, S.M., Haven, E., Hiesmayr, B., Jaeger, G., Khrennikov, A., Larsson, J.A.: Foundations of Probability and Physics-6. Series Conference Proceedings, vol. 1424. American Institute of Physics, Melville (2012)
35. Pothos, E.M., Busemeyer, J.R.: A quantum probability explanation for violation of rational decision theory. Proc. R. Soc. B **276**(1165), 2171–2178 (2009)
36. Asano, M., Ohya, M., Khrennikov, A.: Quantum-like model for decision making process in two players game. Found. Phys. **41**(3), 538–548 (2010)
37. Asano, M., Ohya, M., Tanaka, Y., Basieva, I., Khrennikov, A.: Quantum-like model of brain's functioning: decision making from decoherence. J. Theor. Biol. **281**(1), 56–64 (2011)
38. Asano, M., Basieva, I., Khrennikov, A., Ohya, M., Tanaka, Y.: Quantum-like generalization of the Bayesian updating scheme for objective and subjective mental uncertainties. J. Math. Psychol. **56**(3), 168–175 (2012)
39. Asano, M., Basieva, I., Khrennikov, A., Ohya, M., Tanaka, Y.: Quantum-like representation of irrational inference. In: Busemeyer, J.R., Dubois, F., Lambert-Mogiliansky, A., Melucci, M.

(eds.) Quantum Interaction 6th International Symposium (QI-2012). Lecture Notes in Computer Science, vol. 7620, pp. 138–147 (2012)

40. Asano, M., Basieva, I., Khrennikov, A., Ohya, M., Tanaka, Y., Yamato, I.: Adaptive dynamics and its application to context dependent systems breaking the classical probability law. In: Busemeyer, J.R., Dubois, F., Lambert-Mogiliansky, A., Melucci, M. (eds.) Quantum Interaction 6th International Symposium (QI-2012). Lecture Notes in Computer Science, vol. 7620, pp. 160–171 (2012)
41. Asano, M., Basieva, I., Khrennikov, A., Ohya, M., Yamato, I.: Non-Kolmogorovian approach to the context-dependent systems breaking the classical probability law. Found. Phys. **43**(7), 895–911 (2013)
42. Accardi, L., Ohya, M.: Compound channels, transition expectations, and liftings. Appl. Math. Optim. **39**, 33–59 (1999)
43. Asano, M., Ohya, M., Tanaka, Y., Khrennikov, A., Basieva, I.: Quantum-like representation of Bayesian updating. In: Proceedings of the International Conference Advances in Quantum Theory, vol. 1327, pp. 57–62. American Institute of Physics (2011)
44. Smith, R.A.: A Compleat System of Opticks. Cambridge, Published by the author (1738)
45. Schröder, H.: Ueber eine optische Inversion bei Betrachtung verkehrter, durch optische Vorrichtung entworfener physischer Bilder. Annalen der Physik **181**(10), 298–311 (1858)
46. Atmanspacher, H., Filk, T., Röme, H.: Complementarity in Bistable Perception, Recasting Reality, pp. 135–150. Springer, Berlin (2009)
47. Leggett, A., Garg, A.: Quantum mechanics versus macroscopic realism: is the flux there when nobody looks? Phys. Rev. Lett. **54**, 857–860 (1985)
48. Bell, J.: Speakable and Unspeakable in Quantum Mechanics. Cambridge University Press, Cambridge (1987)
49. Khrennikov, A.: Contextual Approach to Quantum Formalism. Springer, Berlin (2009)
50. Conte, E., Khrennikov, A., Todarello, O., Federici, A., Zbilut, J.P.: A preliminary experimental verification on the possibility of Bell inequality violation in mental states. Neuroquantology **6**, 214–221 (2008)
51. Dzhafarov, E.N., Kujala, J.V.: Quantum entanglement and the issue of selective influences in psychology: an overview. Lect. Notes Comput. Sci. **7620**, 184–195 (2012)
52. Atmanspacher, H., Filk, T.: Temporal nonlocality in bistable perception. In: Khrennikov, A., Atmanspacher, H., Migdall, A., Polyakov, S. (eds.) Quantum Theory: Reconsiderations of Foundations-6. Special Section: Quantum-like Decision Making: From Biology to Behavioral Economics, AIP Conference Proceedings, vol. 1508, pp. 79–88 (2012)
53. Atmanspacher, H., Filk, T., Römer, H.: Weak quantum theory: formal framework and selected applications. In: Adenier, G., Khrennikov, A., Nieuwenhuizen, T.M. (eds.) Quantum Theory: Reconsideration of Foundations, vol. 3, pp. 34–46. American Institute of Physics, New York (2006)
54. Kolmogoroff, A. N.: Grundbegriffe der Wahrscheinlichkeitsrechnung. Springer, Berlin (1933). English translation: Kolmogorov, A.N.: Foundations of Theory of Probability. Chelsea Publishing Company, New York (1956)
55. Acacio de Barros, J.: Joint probabilities and quantum cognition. In: Proceedings of the International Conference on Quantum Theory: Reconsiderations of Foundations-6, Växjö, Sweden, 11–14 June (2012)
56. Goggin, M.E., Almeida, M.P., Barbieri, M., Lanyon, B.P., O'Briend, J.L., White, A.G., Pryde, G.J.: Violation of the Leggett-Garg inequality with weak measurements of photons. Proc. Natl. Acad. Sci. **108**(4), 1256–1261 (2010)

Chapter 7
Operational Approach to Modern Theory of Evolution

Abstract In this chapter we discuss the possibility to apply the quantum formalism to model biological evolution, especially at the cellular level—genetic and epigenetic evolutions. We start with an extended historical introduction which can be useful for non-biologists, physicists, mathematicians, psychologists, sociologists. In particular, detailed comparative analysis of Darwinism and Lamarckism is presented. We discuss the process of elimination of the Lamarckian viewpoint from evolutionary biology and its sudden reappearance in theory of epigenetic evolution. One of the main messages of this chapter is that the ideology of quantum adaptive dynamics (and, in particular, of theory of open quantum systems) matches perfectly with both approaches (Darwinism and Lamarckism) and it can be used to unify them in a new general theory, theory of quantum adaptive evolution.

Keywords Epigenetic cellular evolution · Lamarckism · Darwinism · Open quantum systems · Modeling of environment · Adaptive antivirus immunity · Evolutionary jumps · Punctuated equilibrium · Graduation

7.1 Lamarckism

Lamarck was one of the first who presented evolution in biology as a scientific theory [1]. One of the basic ideas of J.B. Lamarck is that an organism can pass on characteristics that it acquired during its lifetime to its offspring. According to Lamarck, evolution is fundamentally *adaptive* in its nature. He emphasized the role of so to say an *adaptive force* in evolution:

Using the modern language, in any bio-system, its subsystems (organs in Lamarck's terminology) communicate actively with each other to inform about their activity and matching with the environmental pressure. This information is transformed even to germ cells which transform the results of adaptation to the next generation.

M. Asano et al., *Quantum Adaptivity in Biology: From Genetics to Cognition*,
DOI 10.1007/978-94-017-9819-8_7

In the framework of the present book this viewpoint is very important, since it matches well with the idea of the authors that the dynamics of the QL-states[1] of living systems is mathematically described by the theory of adaptive dynamical systems (Chap. 2) and, in particular, the theory of open quantum systems.

However, Lamarckian theory of evolution is not reduced to adaptive inheritance. According to Lamarck, the adaptive dynamics increases information and, consequently, biological *complexity* of bio-systems. This is so to say *complexifying force.*

Our adaptive dynamics approach models the "adaptive force". However, the place of the "complexifying force" in theory of adaptive and, in particular, open quantum systems is not completely clear.

To match with the Lamarck model of complexity increasing evolution, we have to find a "natural class" of QL adaptive dynamical systems, in the simplest case a class of quantum Markov dynamical systems, which increases the complexity of states.

We remark that, in spite of the world-wide success of Darwinism, ideas of Lamarck, especially the ideas about inheritance of the markers of adaptivity, were supported by many authoritative biologists. We can mention I. Pavlov who claimed that in his experimental studies on *conditioned reflex* he found out that this reflex can be inherited and thus the reaction to conditioning is improved from one generation to another. Unfortunately, the validity of his claim has never been tested by other researchers (although it seems to be a routine test).

Lamarckism combined with genetics is known as *neo-Lamarckism*: the environmental pressure induces adaptive mutations in genes which are transferred to the offspring. There exist a lot of experimental evidence both for and against neo-Darwinism (Sect. 7.3). Recently, the ideas of neo-Lamarckism have been extended from genome to *epigenome*. Opposite to genetics, in epigenetics the presence of the Lamarckian component in the cellular evolution is well recognized in the community of evolutionary biologists (Sect. 7.3).

7.2 Darwinism

7.2.1 Survival of the Fittest

The modern theory of biological evolution is fundamentally based on Darwin's book "The Origin of Species by Means of Natural Selection" [2]. One of the main results of this book was the conclusion (including numerous examples) that in the biological world evolution really takes place. All organisms without exception have descended, with modifications, from the common ancestor. This conclusion matched with the general development of the biological concepts before Darwin and, in particular, with Lamarckian views. However, the main novel achievement of Darwin was to

[1] We remind that quantum-like (QL) states are information states representing probabilistic information about possible measurements on bio-systems, including self-measurements.

emphasize the fundamental role of *natural selection* in the evolution of populations of biological organisms. Nowadays Darwinian natural selection is often presented in the form of "*survival of the fittest*". Although this is basically true, Darwin's theory was not so primitive. "Survival of the fittest" is only one of the main factors of Darwinian evolution. In fact, Darwin did not reject completely the (Lamarckian) idea about the inheritance of acquired characteristics, the theory of *pangenesis*.

However, later the role of natural selection as the unique source of evolution was stressed by Wallace [3] and Weismann [4]. They totally rejected the Lamarckian idea of inheritance of acquired characteristics. According to Weismann, germ plasm, or gametes (such as sperm and egg in animals), separate from the somatic cells that go on to make other body tissues at an early stage of the development. Thus (according to Weismann) gametes and somatic cells do not communicate and, hence, environmentally induced changes in somatic cells cannot be transmitted to gametes and these changes cannot be inherited (so called *Weismann's barrier*). We remark that nowadays the crucial role of *cells signaling* in basic cellular processes is well recognized and Weismann's conclusion cannot be considered as justified. However, at the end of 19th century the presence of Weismann's barrier was the deadly argument against the inheritance of acquired characteristics. Darwinism released from the elements of Lamarckism got the name of *neo-Darwinism.* The Wallace-Weismann version of Darwinism was actually the theory of "survival of the fittest."

Nowadays the term neo-Darwinism is used for the theory of evolution driven by natural selection acting on variation produced by *genetic mutation and genetic recombination.* Thus the Darwinian model of evolution liberated from ideas of Lamarckism was combined with genetics. Purely random genetic variations are subject to natural selection under the environmental pressure. Neo-Darwinism is based on the main postulate of molecular biology: In a cell the information flow is possible only in one direction from DNA (RNA) to proteins, from genotype to phenotype; never backward: From phenotype to genotype.

7.2.2 Evolutionary Synthesis/neo-Darwinism

However, the term neo-Darwinism is often used as the equivalent of *evolutionary synthesis.* Here the natural selection is not considered as the sole source of changing genetic variation in biological populations. In particular, pure random *genetic drift* plays an important role in the dynamics of frequencies of alleles (gene variants). (Genetic drift or allelic drift is the change in the frequency of an allele in a population due to purely random sampling, i.e., without the environmental pressure.) In some situations genetic drift is not a mechanism of evolutionary changes less powerful than natural selection driven by the environmental pressure. A number of advanced mathematical models describing the contribution of genetic drift to the process of evolution were elaborated, e.g., the Wright-Fisher model.

The evolutionary synthesis combines various sources of evolution. We point to the following three basic forces of evolution:

ES1 Random mutations exposed to natural selection.
ES2 Genetic drift.
ES3 Ecological factors such as niche occupation and barriers to gene flow.

Evolution is gradual: Small genetic changes regulated by natural selection accumulate over long periods. Discontinuities amongst species (or other taxa) are explained as originating gradually through geographical separation and extinction (however not saltation cf. Sect. 8.2.4). In complete agreement with the views of Wallace and Weismann, the inheritance of acquired characteristics is impossible. We finish our presentation of evolutionary synthesis by the following citation from the book of Futuyma [5], p. 12:

> The major tenets of the evolutionary synthesis, then, were that populations contain genetic variation that arises by random (i.e., not adaptively directed) mutation and recombination; that populations evolve by changes in gene frequency brought about by random genetic drift, gene flow, and especially natural selection; that most adaptive genetic variants have individually slight phenotypic effects so that phenotypic changes are gradual (although some alleles with discrete effects may be advantageous, as in certain color polymorphisms); that diversification comes about by speciation, which normally entails the gradual evolution of reproductive isolation among populations; and that these processes, continued for sufficiently long, give rise to changes of such great magnitude as to warrant the designation of higher taxonomic levels (genera, families, and so forth).

In this book we shall use both terms, neo-Darwinism and evolutionary synthesis, as equivalent. The most important thing for our further consideration is the total liberation of the theory from elements of the Lamarckism.

From **ES1–ES3** and the citation of Futuyma, one can see that the evolutionary dynamics is a combination of many counterparts of totally different origins. How to model the resulting output of the interaction of so many factors? There have been elaborated advanced mathematical models of genetic drift and natural selection and their inter-relation. One can model effects of separation and so on. However, the main problem of modeling synthesis of its counterparts is the impossibility to estimate their contributions. As an example, we can consider the debates on the role of genetic drift. This notion was introduced by Wright [6], who emphasized the role of the sample errors in the processes of evolution. However, he was strongly criticized by Fisher [7], who proposed a number of mathematical models of genetic drift and claimed that, although it affects genetic variation in populations, the magnitude of its effect is small comparing with the effect of natural selection. As the result of Fisher's critique (as well as other critical attacks), genetic drift was considered as a minor contribution to evolution. However, later Kimura [8] coupled the debate on the importance of genetic drift with his *neutral theory of molecular evolution*, which claims that most of the genetic changes are caused by the genetic drift acting on neutral mutations. And recently the role of genetic drift has been criticized by Gillespie [9].

In our opinion, this debate can continue for ever. Contributions of the genetic drift and natural selection may vary essentially depending on context. It is difficult, if at all possible, to estimate relative contributions of these two factors. The same is valid for accounting contributions of other factors of evolution (both known and unknown). Nowadays even the specter of Lamarck walks freely in halls and labs of biological departments of the world's top universities. It declares that the acquired characteristics of cells play a crucial role in cellular evolution, see Koonin and Wolf [10].

7.3 Neo-Lamarckism: CRISPR-Cas System of Adaptive Antivirus Immunity

As pointed out in [10], p. 1: "...massive sequencing of numerous, complete microbial genomes revealed novel evolutionary phenomena, the most fundamental of these being:

(1) pervasive *horizontal gene transfer* (HGT), in large part mediated by viruses and plasmids, that shapes the genomes of archaea and bacteria and call for a radical revision (if not abandonment) of the Tree of Life concept,[2]
(2) Lamarckian-type inheritance that appears to be critical for antivirus defense and other forms of adaptation in prokaryotes,
(3) evolution of evolvability, i.e., dedicated mechanisms for evolution such as vehicles for HGT and stress-induced mutagenesis systems."

Thus these authors also emphasize that evolution (especially for archaea and bacteria) is a multi-factorial process. Experimental studies on evolution of bacteria questioned one of the fundamental principle of neo-Darwinism, namely, the impossibility of specific changes in genome determined by environment. We again cite Koonin and Wolf [10], p. 6: "Recently several genetic phenomena with a distinct Lamarckian flavor have been discovered. Probably, the most striking case is the system of adaptive antivirus immunity, known as CRISPR-Cas (Clustered Regularly Inter-spaced Palindromic Repeats and CRISPR-associated proteins), that is present in most archaea and many bacteria. The CRISPR-Cas system integrates fragments of virus or plasmid DNA into a distinct, repetitive locus in the archaeal or bacterial genome. The transcript of this unique spacer functions as a guide RNA that is incorporated into a specific complex of Cas proteins possessing DNAs activity and directs this complex to the cognate alien DNA (or RNA) molecules that are cleaved and accordingly inactivated. The CRISPR-Cas system is amazingly efficient, with only about 10^{-5} failure rate. This mechanism qualifies CRISPR-Cas as an adaptive immunity system, i.e., immunity system that adapts to a specific infectious agent, a novelty in prokaryotes. Furthermore, the Lamarckian principle of inheritance and evolution is apparent in the mechanism of CRISPR-Cas function. Indeed, this system directly responds to an environmental cue (in this case, foreign DNA) by introducing

[2] The famous sole illustration of the "Origin of Species" shows a Tree of Life (or more precisely, a series of trees presumably depicting the evolution of different divisions of organisms).

a genetic change into the genome that is immediately adaptive with respect to that particular cue."

We also point to some other claims of experimental evidence in favor of (neo-)Lamarckism directly on the genetic level. In a few articles there was claimed that special environments can stimulate mutations in genome exceeding the predictions based on Darwin's model. The first paper in this direction was published by Ryan [11]. His primary aim was to confirm (for a new experimental context) the famous *fluctuation test* which was first performed by Luria and Delbruck [12]. In the literature on foundations of neo-Darwinism, this test is considered as one of the basic genetic confirmations of the neo-Darwinian mechanism of adaptive evolution: First the totally random mutation and then the selection induced by an environment. The main idea of this test can be presented as follows[3]:

> Take a large family of totally independent bacterial cultures and plate them onto separate Petri dishes to grow under the pressure of some special "unfriendly environment". Only those who have special mutation would survive and produce large populations. If mutations are totally random, then one can expect high variation in sizes of final populations. In one culture, a mutant may appear very early and give rise to numerous colonies, in another, it may appear very late and give rise to fewer colonies. As a result, there will be much more fluctuations from one plate to another than in the case of environment induced mutations. Suppose now that mutations appear only as a direct response to the environment. One can also expect some kind of variation, however, the variations from one culture to another would not be very high. And the experiment of Luria and Delbruck demonstrated that so to say "Darwin was right (in fact, Wallace and Weismann), but Lamarck not". The variation in populations was too high to be explained by consistent environment driven mutations in all populations. We also point out that simple probabilistic reasoning implies that in the latter case the probability distribution of variations between populations has to be Poissonian. And the Luria-Delbruck distribution differs non-negligibly from the Poissonian distribution.

Ryan [11] repeated this test (with another environment[4]) and he found that the probability distribution differs essentially from the Luria-Delbruck distribution and it is closer to the Poissonian distribution corresponding to the neo-Lamarckian scenario. This experimental result did not attract much attention; it was practically ignored. However, later an influential paper confirming Ryan's statistics was published in *Nature* by Cairns et al. [14]. Then numerous experiments were performed and a series of papers in good journals was published, see [15–22]. However, whether the obtained statistical data can be explained by cellular natural selection or some contribution of adaptive mutations is present, is still the subject of hot debates. One of the explanations of the data of the "Lamarckian type" is that the pressure from unfriendly environment is considered by cell population as a stress, which increases the total rate of (random) mutations including the environment matching mutations. This explanation was considered as satisfactory by the genetic community

---

[3] Here we follow the presentation from the work [13] of Vasily Ogryzko. He was one of the pioneers in attempts to use quantum *physics* to model cellular evolution.

[4] He used a mutation that would enable the cells to grow on lactose, reverting a previously inactive β-galactosidase gene back to its wild type, Chap. 5.

and (around 2007) the specter of Lamarck disappeared from labs. However, as we have seen, he is back once again: now with the CRISPR-Cas phenomenon.

Finally, we cite Koonin and Wolf once again [10]:

> More generally, recent empirical and theoretical studies of diverse processes of stochastic and deterministic changes in genomes make it clear that evolution is not limited to the basic Darwinian scheme of random variation that is subject to selection. Evolution can be more adequately depicted as a continuum of processes from completely random ones, under the Wrightean modality defined by random variation and random fixation of changes via genetic drift; to the Darwinian modality with ran-dom changes fixed by the deterministic process of selection; to the Lamarckian mode in which both variation and fixation are deterministic.

We can summarize the above considerations as follows:

It is very difficult to split contributions of different components of the evolutionary processes. Therefore it seems really impossible to model evolution through accounting contributions of all possible factors and their interrelation. In this situation we consider the usage of the operational quantum-like approach, cf. Chap. 2, as the only possibility to create a general model of biological evolution. The operational approach makes practically meaningless hot debates about superiority of one factor of evolution over another. Thus, instead of (neo-)Darwinism contra (neo-)Lamarckism or random drift contra natural selection, we shall proceed with the operational quantum(-like) master equation (Chap. 2) unifying all the aforementioned components of the process of evolution. Evolutionary synthesis has to be modeled synthetically, and the theory of open quantum systems and its generalization, adaptive dynamics (Chap. 4), provides such a possibility.

7.4 Evolutionary Jumps; Punctuated Equilibrium and Gradualism

Darwinism as well as the evolutionary synthesis claim that evolution is gradual: Small genetic changes regulated by natural selection accumulate over long periods. From the very beginning this postulate of Darwin was questioned by many evolutionary biologists. The active discussion on *evolutionary jumps* (leaps, saltations, transiliencies) was started by Galton in 1894 [23], who attacked Darwin's theory of evolution through small, incremental steps (see Gillham [24, 25] for the detailed historical presentation).

No variation could "establish itself unless it be of the character of a sport, that is, by a leap from one position of organic stability to another, or as we may phrase it through E-transilient, E-variation." Galton was "unable to conceive the possibility of evolutionary progress except by transiliencies." A transiliency is a saltatory change, a jump or a leap, from one state to another, one race to a new race, one species to a new species. In fact, as was rightly pointed by Gillham, the term saltation (from Latin, saltus, "leap"), a sudden change from one generation to the next, that is large, or very large, in comparison with the usual variation of an organism, would be more appropriate than transiliency.

Galton felt that the small, incremental steps by which natural selection supposedly proceeded would be thwarted by the phenomenon he had discovered, which he called regression (or reversion) to the mean. Hence, Galton believed that evolution must proceed via discontinuous steps.

This debate between Galton's and Darwin's adherents was later transformed into the well known debate between adherents of Mendelianism and biometricians. Since at the beginning Mendel's theory was in visible contradiction with Darwin's continuous evolution, Bateson, one of the most prominent aliens of Galton, see [25], actively used Mendel's laws as supporting evolution by jumps [26]. However, later it became clear that Mendel's approach also can be used to explain continuous changes a la Darwin. Nowadays, although it is commonly accepted that there were no evolutionary jumps, the debate between Galtonists and Darwinists continues. Recently, a form of Galtonism has been reborn in the theory of *punctuated equilibrium*.

Although S. J. Gould and N. Eldredge, the creators of the theory of punctuated equilibrium [27], wanted to distance from evolutionary models based on saltations (mainly because nowadays such evolutionary models are actively used by adherents of creationism in evolution), the theory of *punctuated equilibrium* essentially based on Galton's ideas. By this theory, evolution consists of periods of rapid change implying creation of new species and long periods of *statis*—stability of species. The main distinctions are (a) the emphasis of statis (such periods can take millions of years and during periods of statis the environment can vary essentially without changing species), (b) the duration of the discrete events, creation of new species. By Galton, such events are really instantaneous (one generation), by Gould and Eldredge, "creation" can take 50,000–100,000 years, so that its duration is short (of the spot-type) only in comparison with the duration of the period of statis.

However, in the same way as Galton, Gould and Eldredge claimed [27] that new species are created not via random mutations and then natural selection in the major population, but through speciation in isolated (e.g., geographically) relatively small sub-populations. In such isolated sub-populations profitable changes would not be totally washed by stochasticity of inbreeding in very large populations.

Galtonism as well as the theory of punctuated equilibrium were created to oppose Darwinian and neo-Darwinian viewpoint that evolution is based on small gradual changes in huge populations and that profitable small changes (selected to match with the environmental pressure) during very long periods lead to new species (Although Gould and Eldredge, who were rather opportunistic in the presentation of their views, tried to combine peacefully punctuated equilibrium with gradualism [27].)

Once again we see that it is very difficult to terminate the aforementioned dispute and make the choice between the discrete and continuous versions of evolution. Therefore the operational QL-approach may again be considered as a tool for unification. We shall see, Sect. 8.2.3, that the evolution described by quantum master equation is combined of jumps and continuous drifts. Thus usage of the apparatus of the theory of open quantum(-like) and more generally adaptive systems can provide a synthetic mathematical description of discrete and continuous components of evolutionary changes.

7.5 Epigenetic Lamarckism

In recent years cell biologists have found out that non-genetic variation acquired during the life of an organism can sometimes be passed on to offspring E-a phenomenon known as *epigenetic inheritance*, see, e.g., Russell's monograph [28] (see also Chap. 3). Recently several examples of adaptive mutations in eukaryotes by the *epigenetic mechanism* have been reported, see Jablonka and Raz [29] for a detailed review. Under the influence of the environment, the epigenome structure including DNA methylation and histone modification may change during growth and such changes sometimes can be inherited by the offspring. This is the adaptive mutation and a kind of *neo-Larmarckism* [29].

Everywhere below we shall use the term *epimutation*: A heritable change in gene expression that does not affect the actual base pair sequence of DNA.

We concentrate modeling on *cellular epigenetic inheritance* (CEI). This is a narrower aspect of epigenetic inheritance as discussed in the broad sense (see also Sect. 3.3.3). It refers to epigenetic transmission in sexual or asexual cell lineages, and the unit of this transmission is the cell. Four types of CEI are recognized today [29]:

1. The CEI based on self-sustaining regulatory loops.
2. The CEI based on three-dimensional templating.
3. The chromatin-marking CEI.
4. The RNA-mediated CEI.

These types of CEI are realized in cells with the aid of very different processes whose mechanisms are known only in general (many important details still have to be clarified). Nevertheless, all the aforementioned CEIs are parts of one universal phenomenon, namely, the development of special adaptive features under the pressure of the environment and transmission of these features from a mother cell to daughter cells.

Therefore a perspective to create a universal model of CEI describing all its types in the common framework is very attractive. Of course, such an activity does not contradict to the continuation of intensive studies in cellular system biology aimed to the creation of detailed models for each CEI and their interrelation. We present an operational model, see Sect. 1.2, of CEI, which is applicable to all its possible types (known and even yet unknown). Here the keyword is *adaptive dynamics* (Sect. 1.4). Our aim is to create an adaptive dynamical model of CEI. This model, although it does not describe concrete cellular mechanisms, can be interesting for cellular biology. It presents a general mathematical structure of CEI and justifies the epigenetic mechanism from the viewpoint of the theory of adaptive dynamical systems. In the light of our model the existence of CEI-mechanisms in cells is very natural, cf. Sect. 1.4. Roughly speaking, if CEI were not discovered in experimental studies, it would be a tricky problem to explain its absence. Thus by using the operational approach and ignoring details of cellular processes we acquire knowledge on universal information processes beyond CEI.

References

1. Lamarck, J.-B.: Philosophie Zoologique ou Exposition des Considerations Relatives à l'Histoire Naturelle des Animaux. Zoological Philosophy: Exposition with Regard to the Natural History of Animals. Dentu, Paris (1809)
2. Darwin, Ch.: The Variation of Animals and Plants Under Domestication. John Murray, London (1868)
3. Wallace, A.R.: Darwinism: An Exposition of the Theory of Natural Selection, With Some of Its Applications. Macmillan and Company, New York (1889)
4. Weismann, A.: Essays Upon Heredity, vols. 1, 2. Clarendon Press, Oxford (1889)
5. Futuyma, D.J.: Evolutionary Biology. Sinauer Associates, Sunderland (1986)
6. Wright, S.: The evolution of dominance. Am. Nat. **63**(689), 556–561 (1929)
7. Fisher, R.A.: The Genetical Theory of Natural Selection. Clarendon Press, Oxford (1930)
8. Kimura, M.: Evolutionary rate at the molecular level. Nature **217**(5129), 624–626 (1968)
9. Gillespie, J.H.: Genetic drift in an infinite population. The Pseudohitchhiking Model Genetics **155**(2), 909–919 (2000)
10. Koonin, E.V., Wolf, Y.I.: Evolution of microbes and viruses: a paradigm shift in evolutionary biology? Front. Cell. Infect. Microbiol. **2**, Article number 119 (2012)
11. Ryan, F.J.: Spontaneous mutation in non-dividing bacteria. Genetics **40**(5), 726–38 (1955)
12. Luria, S.E., Delbruck, M.: Mutations of bacteria from virus sensitivity to virus resistance. Genetics **28**(6), 491–511 (1943)
13. Ogryzko, V.V.: On two quantum approaches to adaptive mutations in bacteria. arxiv:ftp/arxiv/papers/0805/0805.4316.pdf
14. Cairns, J., Overbaugh, J., Miller, S.: The origin of mutants. Nature **335**(6186), 142–145 (1988)
15. Foster, P.L.: Adaptive mutation: the uses of adversity. Annu. Rev. Microbiol. **47**, 467–504 (1993)
16. Foster, P.L.: Adaptive mutation: implications for evolution. Bioessays **22**(12), 1067–1074 (2000)
17. Hall, B.G.: Is the occurrence of some spontaneous mutations directed by environmental challenges? New Biol. **3**(8), 729–733 (1991)
18. Hall, B.G.: Adaptive mutagenesis: a process that generates almost exclusively beneficial mutations. Genetica **102–103**(1–6), 109–125 (1998)
19. Kugelberg, E., Kofoid, E., Reams, A.B., Andersson, D.I., Roth, J.R.: Multiple pathways of selected gene amplification during adaptive mutation. Proc. Natl. Acad. Sci. USA **103**(46), 17319–17324 (2006)
20. Pettersson, M.E., Roth, J.R., Berg, O.G.: The amplification model for adaptive mutation: simulations and analysis. Genetics **169**(2), 1105–1115 (2005)
21. Roth, J.R., Kugelberg, E., Reams, A.B., Kofoid, E., Andersson, D.I.: Origin of mutations under selection: the adaptive mutation controversy. Annu. Rev. Microbiol. **60**, 477–501 (2006)
22. Stumpf, J.D., Poteete, A.R., Foster, P.L.: Amplification of lac cannot account for adaptive mutation to Lac+ in *Escherichia coli*. J. Bacteriol. **189**(6), 2291–2299 (2007)
23. Galton, F.: Discontinuity in evolution. Mind (n.s.) **3**, 362–372 (1894)
24. Gillham, N.W.: A Life of Sir Francis Galton. Oxford University Press, Oxford (2001)
25. Gillham, N.W.: Evolution by jumps: Francis Galton and William Bateson and the Mechanism of evolutionary change. Genetics **159**, 1383–1392 (2001)
26. Bateson, W.: Materials for the Study of Variation Treated With Especial Regard to Discontinuity in the Origin of Species. Macmillan, London (1894)
27. Gould, S.J., Eldredge, N.: Punctuated equilibrium comes of age. Nature **366**, 223–227 (1993)
28. Russell, P.J.: iGenetics. Pearson Benjamin Cummings, San Francisco (2002)
29. Jablonka, E., Raz, G.: Transgenerational epigenetic inheritance: prevalence, mechanisms, and implications for the study of heredity and evolution. Q. Rev. Biol. **84**, 131–176 (2009)

Chapter 8
Epigenetic Evolution and Theory of Open Quantum Systems: Unifying Lamarckism and Darwinism

Abstract This chapter is devoted to a model of the epigenetic cellular evolution based on the mathematical formalism of open quantum systems. We emphasize that, although in this book we restrict our QL-modeling to the epigenetic evolution, it is clear that the structure of the model allows it to be extended to describe evolution of biological organisms in general. We restrict the model to epigenetics, since here we can use a closer analogy with quantum mechanics and mimic behavior of a cell as behavior of a quantum particle. In general, we have to take into account cell death (annihilation in quantum terminology) and birth (creation). Mathematically, such a model is more complicated. One of the basic quantum information constructions used in this chapter is entanglement of quantum states. Our evolutionary model is based on representation of the epigenetic state of a cell as entanglement of various epigenetic markers.

Keywords Epigenetic markers · Epigenetic evolution · Darwinism · Epigenetic Lamarckism · Open quantum systems · Quantum master equation · Entanglement · Superstong quantum(-like) correlations for epigenetic markers · Evolutionary jump version of quantum jump · Punctuated equilibrium · Graduation · Cells signaling

This chapter is devoted to a model of the epigenetic cellular evolution based on the mathematical formalism of open quantum systems. We emphasize that, although in this book we restrict our QL-modeling to the epigenetic evolution, it is clear that the structure of the model allows it to be extended to describe evolution of biological organisms in general. We restrict the model to epigenetics, since here we can use a closer analogy with quantum mechanics and mimic behavior of a cell as behavior of a quantum particle. In general, we have to take into account cell death (annihilation in quantum terminology) and birth (creation). Mathematically, such a model is more complicated.

What kind of the mathematical apparatus can be used to model the adaptive dynamics operationally? As pointed out in Sect. 2.4, an important class of adaptive dynamical systems can be described by the apparatus of the theory of *open quantum systems*. Now we apply this theory to the epigenetic dynamics of a cell. In accordance with the theory of open quantum systems, dynamics of the epigenetic state of a

M. Asano et al., *Quantum Adaptivity in Biology: From Genetics to Cognition*,
DOI 10.1007/978-94-017-9819-8_8

cell is approximately described by the *quantum master equation*, GKSL-equation, see Eq. (2.50) (Sect. 2.4).

Another key notion of our model of epigenetic evolution is *entanglement*, Sect. 2.5.1. (In the process of evolution the epigenetic state becomes *entangled* with the environment). Hence, we shall consider the environment-trace dynamics of the compound system (a cell unified with the environment). This trace dynamics is very complex and therefore one typically considers its (quantum) Markovian approximation given by GKSL-equation. (Non-Markovian quantum dynamics can also be useful for applications to cell biology, but they are based on essentially more advanced mathematics. Hence, as we did in cognitive science, we prefer to apply the approximate Markovian dynamics for simplicity reasons.)

In our model the same dynamical (GKSL-)equation not only describes epimutations (induced by the environment), but also the process of selection of these epimutations leading to the formation of a stable phenotype (Sect. 8.2.3). Our QL-model *unifies neo-Darwinism (Evolutionary Synthesis) and neo-Lamarckism,* but on the epigenetic level, cf. Jablonka and Raz [1]; see also Koonin and Wolf [35] for unification of (neo-)Darwinism and (neo-)Lamarckism on the genetic level. This is a model of adaptive epimutations and "natural selection" in a single living cell.[1] Both Darwinism and Lamarckism contribute to our model in modified forms. The environment driven selection is performed on the molecular level in a single cell. In the process of such an evolution the selection is not of the type "to survive or to die" (as in the orthodox neo-Darwinistic approach), but each cell selects consistently the structure of epigenetic mutations to approach the complete matching with the environment.

Hence, our main claim is the following:

The epigenetic evolution is neither pure Darwinian nor pure Lamarckian, but it combines basic features of both models of evolution.

In complete accordance with the ideology of quantum mechanics, a cell and an environment evolve as a unified system. The state of such a compound system cannot be reduced to the states of its subsystems; it is the state of *entanglement.* A cell loses its individuality and becomes a part of a compound system.

In our model, at the beginning of the interaction with an environment the (epigenetic) state of a cell is characterized by a *high degree of uncertainty about possible epigenetic changes*, which can be generated via the coupling with an environment. The quantum master equation describes the process of resolution of this state of uncertainty and approaching the complete matching with the environment. This process can be considered as *decoherence of cell's state* through the interaction with an environment (cf. Sect. 2.4). As a result, cell epigenetic state loses its fundamentally quantum(-like) feature, the superposition of a few alternatives, and the final situation[2] can be described by the classical probability theory, see [2–4] and Chap. 4 for a similar cognitive model for decision making.

---

[1] Such kind of natural selection is performed not on the cellular, but on the molecular level.

[2] Mathematically, it is characterized by approaching a steady state solution of the quantum master equation (Sect. 2.4).

Once again, we state that we work in the framework of quantum bio-informatics and not quantum biology. Our QL-model should be distinguished from real quantum models in cellular biology. Such models are based on "*quantum reductionism*", i.e., the reduction of cell behavior to the behavior of quantum particles inside a cell, e.g., Ogryzko [5, 6], McFadden and Al-Khalili [7], McFadden [8]. In particular, in these works superpositions of cell states are induced by real quantum superpositions of states of quantum particles inside the cell, e.g., some protons in DNA. These models of evolution were strongly criticized in the framework of quantum biology since quantum physics does not support speculations on the possibility to create a macroscopic superposition and entanglement in cells as physical systems and, moreover, to preserve them for sufficiently long time, see Donald [9] for a very detailed presentation of critical arguments. Even the idea that a cell interacting with an environment can process information in a same vein as quantum computer has no physical support (at least at the moment). *Unitary evolution characterizing quantum computations is a feature of isolated physical systems*, see [9] for critique of usage of the standard quantum computational scheme in biology.

8.1 Information Interpretation

We state once again that our QL-description is simply an operational description of information processes in cells. Even N. Bohr characterized *quantum mechanics as an operational approach to physics* of micro-world: quantum mechanics describes only statistics of measurements (Sect. 1.2). We extended Bohr's viewpoint to cell biology.

The operational approach to the quantum formalism can be completed by the *information interpretation*, e.g., Fuchs [10] (see also [11, 12]) and Zeilinger [13]. Nowadays this interpretation is very popular. By this interpretation, quantum mechanics describes neither waves nor particles. Its formalism describes information about expected results of measurements for quantum systems.[3] We also point to the works of Balazs [14] and Karafyllidis [15], which can be interpreted as genetic models in the QL information framework.

8.1.1 Schrödinger Cat Meets "Schrödinger E. coli *Bacteria"*

One of the basic examples illustrating the need of the information interpretation of the wave function (the quantum state) was proposed by E. Schrödinger. It is known as *Schrödinger's cat.*

[3] By using the information interpretation of the quantum formalism we need not keep to quantum physical reductionism as, e.g., Ogryzko [5, 6], McFadden and Al-Khalili [7], McFadden [8].

Schrödinger's cat itself may know its real state, alive or dead. But we, observers, do not know this state. Therefore, we think it is at the superposed state of alive or dead. In this sense, even the *E. coli* in glucose/lactose experiments is at the superposition of the lactose oriented or glucose oriented states (Chap. 5). *E. coli* itself may know its state, but we do not know it before "asking it the question" expressed in the experiment on growing on lactose/glucose and assaying its β-galactosidase. Hence, before the assay, we should say that *E. coli* is at the state of superposition, see again Chap. 5.

We have already pointed that the mathematical formalism of quantum mechanics delivers only statistical predictions to an observer. The information encoded in the wave function is about *probabilities* of possible results of measurements. This formalism cannot say anything about a possible result of measurement for a specific single system. Therefore our model of the QL-evolution is of fundamentally probabilistic nature. It describes the evolution of probabilities for genes' expressions in large ensembles of cells. What happens in a single cell is the subject of system biology.

8.2 Dynamics of Cell Epigenetic State in the Process of Interaction with an Environment

In this section, we shall present a formal description of dynamics of the epigenetic states of cells interacting with an environment. In Sect. 8.3 this scheme will be concretized.

Denote the space of QL-states of cell epigenome by the symbol $\mathscr{H}_{\text{epi}}$. These states represent statistical information about possible observations on phenotype changes.[4] The structure of this space will be discussed in Sect. 8.3. For the moment, it is simply a complex Hilbert space. In an approximate model we can assume that $\mathscr{H}_{\text{epi}}$ is finite dimensional, i.e., it can be represented as the Cartesian product of a few copies of the field of complex numbers $\mathbb{C}$. Thus $\mathscr{H}_{\text{epi}} = \mathbb{C}^N$. The space of QL-states of the environment is denoted by $\mathscr{H}_{\text{env}}$. This state can have huge dimension (and even be infinite-dimensional). However, this is not a problem. Since, finally, we shall be interested only in the dynamics of cell epigenetic state, the degrees of freedom of the environment will be excluded from the direct consideration by tracing with respect to the space $\mathscr{H}_{\text{env}}$.

The state space of the compound system is the tensor product $\mathscr{H}_{\text{epi}} \otimes \mathscr{H}_{\text{env}}$, see Sect. 2.5. However, as mentioned above, we shall not operate in this tensor space; the dynamics will be restricted to the space $\mathscr{H}_{\text{epi}}$, the contribution of dynamics in $\mathscr{H}_{\text{env}}$ will be embedded into the (operator) coefficients of the quantum master equation.

[4] As was already emphasized, our model treats all possible sources of CEI and their interrelations in the universal framework by extracting fundamental features of information processes leading to CEI.

8.2.1 Epigenetic Evolution from Quantum Master Equation

Our proposal is to use the machinery of the theory of *open quantum systems* and to describe the dynamics of the epigenetic QL-state by using the quantum master equation (the GKSL-equation), cf. with QL-cognitive modeling [2–4] and Chap. 6. This equation can be used to describe transitions from states of uncertainty given by QL-superpositions to classical probability distributions. Hence, such an equation cannot be an equation with respect to the quantum state represented as a *vector* belonging to the complex Hilbert space (normalized by one)-a *pure state*. Such vectors represent superpositions of possibilities, which have to disappear at the end of the epigenetic evolution. We have to use more general quantum states represented by *density operators*. Both the purely quantum superposition describing uncertainty and the final classical probability distribution can be represented by density operators (Sect. 2.2).

In the quantum Markovian approximation the dynamics of the state of a system interacting with an environment is described by the GKSL-equation, see (2.50):

$$\gamma \frac{d\rho_{\text{epi}}}{dt} = -i[H, \rho_{\text{epi}}(t)] + \mathcal{W}\rho_{\text{epi}}(t),\ \rho_{\text{epi}}(0) = \rho^0_{\text{epi}}, \tag{8.1}$$

where H is a Hermitian operator determining the internal dynamics of epimutational changes in cells, which are isolated from the environmental pressure ("cell's Hamiltonian"), and the linear operator $\mathcal{W}$ describes the environmental pressure. In general, opposite to H, the operator $\mathcal{W}$ has a complex mathematical structure. It has such a form that starting with a density operator ρ^0_{epi} we shall get density operators at all instances of time. For the time being, the specific structure of $\mathcal{W}$ is not important for us, see Sect. 8.2.3 for mathematical details. Biologically, this operator is determined by the properties of the environment, including the initial state of the environment. Here γ is the time scale constant, which determines the temporal dimension of the epigenetic evolution. By using such a scaling factor of the dimension of time, we are able to proceed with dimensionless Hamiltonian H and the environmental operator $\mathcal{W}$.

For a very general class of GKSL-equations, the environmental operator $\mathcal{W}$ drives (in the limit $t \to \infty$) the epigenetic state of an ensemble of cells, $\rho_{\text{epi}}(t)$, to the *steady solution*: $\rho_{\text{epi}}(t) \to \rho_{\text{epi};\text{st}}$. Typically, the uncertainty (in the form of superposition) is eliminated from the asymptotic state $\rho_{\text{epi};\text{st}}$.

In our QL-model such a steady state is considered as the result of the epigenetic evolution in the environment (mathematically represented by the operator $\mathcal{W}$). The limiting probability distribution $\rho_{\text{epi};\text{st}}$ describes the probability distribution of epimutations, which took place in a cell population as a consequence of interaction with the environment. The internal uncertainty whether to (epi)mutate or not mutate, was resolved and a stable *phenotype* was created.

This QL-model can be considered as a sort of neo-Lamarckism; *epigenetic Lamarckism.* However, (surprisingly) it also contains an important element of the *epigenetic Darwinism* (natural selection of epigenetic markers inside a cell). As we shall show, Sect. 8.2.3, the environmental operator $\mathcal{W}$ contains (in general, numerous) jump and continuous drift components. The latter can be considered as modeling the natural selection of epimutations.

This process of the transition of the initially superposed state into the classical statistical mixture of a few possibilities is called *decoherence.* Hence, in our model the epigenetic evolution can be considered as the process of decoherence.

Finally, we remark that under natural restrictions, a selection operator produces the same steady state for all possible initial states. Therefore the variety of internal epigenetic states produced by the Schrödinger's dynamics before the environment started to play a crucial role is transformed into the same steady state, the fixed phenotype (Sect. 8.3.1).

8.2.2 On Applicability of Quantum Master Equation to Description of dynamics of Epigenome

The quantum Markovian dynamics (8.1) is derived (see, e.g., [16, 17]) under the requirement that the reaction of the environment to a system is negligibly small. This requirement is natural in modeling of the epigenetic evolution. We can assume that a cell cannot change essentially the structure of the environment. Another condition for derivation of Eq. (8.1) is the factorization of the initial state of the compound system: The cell in combination with the environment. In quantum terms this means that at the beginning the states of cells and the environment were not *entangled.* In cell biological terms it means that before being under the influence of the environment the population of cells and the environment were not correlated, i.e., the previous evolution of cells was independent from this specific environment.

This special form of the open quantum dynamics is derived under the assumption of Markovness. In the cell biological framework this assumption has the following form. A cell does not "remember" a long chain of interactions with the environment; its state at the instant of time $t + \delta t$, $\psi_C(t + \delta t)$, where δt is a very small (mathematically infinitely small) interval, is determined by its state at the instant of time t, $\psi_C(t)$, and not by the family of its states in previous instants of time, i.e., $\{\psi_C(s) : s < t\}$. This is the most questionable assumption. We cannot exclude the presence of long term memory effects in the epigenome of a cell interacting with an environment. As well as in physics, we treat the Markovian condition as an approximate condition, i.e., long term memory effects may be present in a cell, but they are sufficiently weak to justify the approximation in use.

8.2.3 Evolutionary Jumps as Quantum-Like Jumps

This section is more complicated mathematically than other sections. In principle, the reader can jump directly to the summary at the end of this section. To enlighten the structure of the GKSL-evolution, we have to discuss the form of the "environment-operator" $\mathscr{W}$ in (8.1). In the finite dimensional case it can be represented as the finite sum, see, e.g., [16, 17]:

$$\mathscr{W}\rho = \sum_j \Big(L_j \rho L_j^* - \frac{1}{2}\{L_j^* L_j, \rho\} \Big), \tag{8.2}$$

where L_j are traceless operators (Lindblad operators) and $\{A, B\} = AB + BA$ denotes the anticommutator of two operators. Consider the dynamics driven by a single Lindblad operator L, see Rooney et al. [18] for details; we set the time scaling $\gamma = 1$:

$$\frac{d\rho}{dt} = L\rho L^* - \frac{1}{2}\{L^* L, \rho\}. \tag{8.3}$$

(We also set $H = 0$. We know that the corresponding H-dynamics, the Schrödinger dynamics, is reduced to fluctuations. Now we are only interested in the impact of the environment).

Hence, for a very small interval of time δt, the state is changed as

$$\rho \to \rho - \frac{1}{2}\{L^* L, \rho\}\delta t + L\rho L^* \delta t + o(\delta t). \tag{8.4}$$

This expression can be written as

$$\rho \to V_0 \rho V_0^* + V_1 \rho V_1^*, \tag{8.5}$$

where V_1 is simply equal to $L\sqrt{\delta t}$ and $V_0 = I - \frac{1}{2} L^* L \delta t$.

Now consider the actions of these operators to a pure state $|\psi\rangle$. We remind that the corresponding density operator is the operator of the orthogonal projection to the state $|\psi\rangle : \rho_\psi = |\psi\rangle\langle\psi|$. The pure state vector is always normalized.

We start with the V_1-action. First, we remark that $L|\psi\rangle\langle\psi|L^* = |L\psi\rangle\langle L\psi|$. Hence, under the V_1 action the original state $|\psi\rangle$ jumps to the state

$$\frac{1}{\| |L\psi\rangle \|} |L\psi\rangle \tag{8.6}$$

(where the denominator is the normalization constant) with the probability

$$P_{\text{jump}} = \| |L\psi\rangle \|^2 \delta t = \langle\psi| L^* L |\psi\rangle \delta t. \tag{8.7}$$

We regard this action as a *jump*, since for $\delta t \to 0$ the output state does not approach the input state $|\psi\rangle$. The probability of no jump is given by $P_{\text{no jump}} = 1 - \langle\psi|L^*L|\psi\rangle\delta t$. However, the absence of a jump does not imply that the input state is preserved. The branch without jump evolves as $|\psi\rangle \to |\psi\rangle - \frac{1}{2}L^*L|\psi\rangle\delta t$ with the corresponding normalization. We call this branch the *drift-type evolution*, since the output state approaches the input state for $\delta t \to 0$.

Thus, in the QL-model the environment driven evolution can be considered as a branching process with *"evolutionary jumps"* ("quantum jumps") [5] and *continuous drift-evolution.* We shall study this problem in more detail in Sect. 8.3.1, where the simplest model of epigenetic mutation in a single gene will be considered.

The sum in representation (8.2) of the environment operator can contain a few terms. The number of terms can be very large, it grows as $N^2 - 1$, where N is the dimension of the state space. In our model the dimension of the epigenetic state space grows as $N = 2^n$, where n is the number of epigenetic markers under consideration, Sects. 8.4 and 8.5. Therefore, in general, the QL evolution is a combination of about 2^n quantum jumps to the states determined by the operators L_j and the corresponding drifts. This is a branching process of great complexity. The epigenetic state jumps in different directions (determined by the environmental operators L_j), outputs of jumps form superpositions; if no L_j-jump occurs, the state deforms continuously, and these continuous deformations are superposed with the superposition of jumps. At the next step, the state directions of jumps and drifts are randomly changed…

8.2.4 On a Quantum-Like Model of Evolution: Dynamics Through Combination of Jumps and Continuous Drifts

In our QL model we cannot escape consideration of evolutionary jumps, see (8.6). It is impossible to reduce the dynamics (8.5) to just its first component, the continuous drift $\rho \to V_0\rho V_0^*$. The law of conservation of probability would be violated. On the other hand, as was already pointed out, the absence of a jump does not imply the stationarity of the state. It has to drift continuously and permanently. Thus by the QL model of evolution of *"open biological systems"* evolutionary jumps ("salutations") are indivisibly coupled with continuous drifts. The first one can be considered as the Galtonian component (see Sect. 7.4) and the second one as the Darwinian component (Ql representation of gradualism) of evolution. Thus in the mathematical framework of the theory of open quantum(-like) systems Galtonism is much closer to Darwinism than it can be imagined. Moreover, in our model Lamarckism is realized as the combination of Galtonism and Darwinism. (We stress that we model cellular evolution.)

---

[5] Quantum jump (leap) is a jump of an electron from one quantum state to another within an atom. Quantum jumps were invented by Einstein, who postulated that electrons in an atom can absorb and emit electromagnetic energy only by discrete portions, which later were called photons. Thus, opposite to classical systems, electron energy cannot change continuously.

8.3 Dynamics of a Single Epimutation of the Chromatin-Marking Type

In this section as well as in Sect. 8.4 we restrict our consideration to epimutation of the chromatin-marking type (Chap. 3). This special case has illustrative advantages: epimutations of this type can be directly coupled with physical carriers, genes, to which DNA methylations and histone modifications can be coupled. Hence, in the same way as in quantum mechanics we can couple a quantum(-like) state with the corresponding physical system, the gene. Of course, our approach is applicable to all the four types of epimutations, which were discussed in the review of Jablonka and Raz [1], see introduction to this chapter. We shall consider the general situation in Sect. 8.5.

Consider the concrete gene g in cell genome. Suppose that this cell interacts with an environment in such a way that some type of epigenetic mutation, say μ, in g can happen. This epimutation changes the level of expression of g.

Ignoring the presence of other genes and corresponding gene expressions we can model the μ-mutation by considering simply the two dimensional state space H_{epi} (qubit space). States of no mutation and mutation are represented by two orthogonal vectors $|0\rangle$ and $|1\rangle$. Hence, a (pure) QL-state can be represented as the superposition

$$|\psi_{\text{epi}}\rangle = c_0|0\rangle + c_1|1\rangle, \tag{8.8}$$

where $c_0, c_1 \in \mathbb{C}$, $|c_0|^2 + |c_1|^2 = 1$. As was remarked, the quantum master equation does not preserve pure states, so sooner or later superposition (8.8) will be transferred into the statistical mixture given by a density matrix. We remark, see the equality (2.31), that in terms of density matrices the pure state (8.8) can be written as

$$\rho_{\text{epi}} = \begin{pmatrix} |c_0|^2 & c_0\bar{c_1} \\ \bar{c_0}c_1 & |c_1|^2 \end{pmatrix}. \tag{8.9}$$

Thus nontrivial superposition is characterized by the presence of nonzero off-diagonal terms. We remark that the absolute value of the off-diagonal terms is maximal and equals 1/2 for the uniform superposition $|\psi_{\text{epi}}\rangle = \frac{1}{\sqrt{2}}(|0\rangle + |1\rangle)$, representing the maximal uncertainty. The dynamics (2.50) suppresses the off-diagonal terms and, finally, a diagonal density matrix (steady state) arises,

$$\rho_{\text{st}} = \begin{pmatrix} \rho_{00;\text{st}} & 0 \\ 0 & \rho_{11;\text{st}} \end{pmatrix}. \tag{8.10}$$

Its elements $\rho_{00;\text{st}}$ and $\rho_{11;\text{st}}$ give probabilities of the events: "No μ-epimutation" and "μ-epimutation". Thus in a large population of cells, say M cells, $M >> 1$, the number of, e.g., cells with mutation is given (approximately) by $N_m \approx \rho_{11;\text{st}}M$. The limiting QL-state (represented by the diagonal matrix (8.10) approached the stability with respect to the influence of this (concrete) environment. We remark that mathematically a population needs infinite time to stabilize completely to the steady state. Therefore in reality one can expect fluctuations (of decreasing amplitude) on a finite interval of time.

We remark that under a special interrelation between operators H and $\mathcal{W}$ the stabilization is achieved with the state ρ_{st} such that $\rho_{11;st} >> \rho_{00;st}$. In such a case the epimutation μ spreads practically to the whole population and, moreover, it will be inherited. Thus the quantum master equation is sufficiently general to represent (on the epigenetic level) the regime similar to one represented by *Fisher's equation* [19] that was used to describe the spreading of biological populations. The main distinguishing feature of the epigenetic situation is that the epimutation spreads in a single generation of cells and then it is inherited by the next generation.

8.3.1 Entropy Decreasing Evolution

As was already pointed out, it is possible to construct QL dynamics such that *starting with any state the trajectory will stabilize to the same pure state* (see Chap. 4 for cognitive applications), e.g., $\rho_{st} = \pi_1 = |1\rangle\langle 1|$. Even if at the beginning only a few cells were (epi)mutated, finally, cells will mutate with the unit probability. By moving from a mixed state, e.g., from the state given by the diagonal matrix with equal elements, to the pure state $|1\rangle$,

$$\rho_0 = \begin{pmatrix} 1/2 & 0 \\ 0 & 1/2 \end{pmatrix} \rightarrow \rho_{st} = \begin{pmatrix} 0 & 0 \\ 0 & 1 \end{pmatrix}, \tag{8.11}$$

the von Neumann (quantum) *entropy decreases*. We emphasize the role of environment in such an evolution.

Consider now the environment operator $\mathcal{W}$ based on a single Lindblad operator, see (8.2), acting in a qubit space and having the form

$$L = \sqrt{p}|1\rangle\langle 0|, \tag{8.12}$$

i.e., $L|\psi\rangle = \sqrt{p}\langle 0|\psi\rangle|1\rangle$. Thus during a small time interval δt, each state $|\psi\rangle$ can jump only towards the same state $|1\rangle$, see (8.6). For the pure state $|\psi\rangle = c_0|0\rangle + c_1|1\rangle$, the probability of such a jump is equal to $P_{jump} = p|c_0|^2\delta t$.[6]

In such a process all evolutionary jumps are oriented towards mutation (the concrete mutation under consideration); cell state cannot jump back by eliminating this mutation. However, the evolution is not reduced to jumps. There is also a continuous evolutionary drift, which changes the probability of a jump.

---

[6] As in Sect. 8.2.3, we proceed under the assumption that the time scale constant γ was set as $\gamma = 1$. If we take γ into account, then the formula for the jump-probability takes the form: $P_{jump} = p|c_0|^2 \frac{\delta t}{\gamma}$. Hence, the smallness of the jump duration is relative to the time scale of evolution.

We now consider the evolutionary drift encoded by the operator $V_0 = I - \frac{1}{2}L^*L = I - \frac{p}{2}\pi_1$, where $\pi_1 = |1\rangle\langle 1|$. The latter operator is simply the orthogonal projector onto the state of mutation $|1\rangle$. For a pure state, this dynamics can be mathematically represented as the dynamics of a vector ("non-normalized state")

$$\frac{d|\phi\rangle}{dt}(t) = -\frac{p}{2}\pi_1|\phi\rangle(t). \tag{8.13}$$

The solution of this linear differential equation is given by $|\phi\rangle(t) = e^{-pt/2}c_{00}|0\rangle + c_{10}|1\rangle$ and the corresponding QL state evolves as

$$|\psi\rangle(t) = \frac{e^{-pt/2}c_{00}|0\rangle + c_{10}|1\rangle}{\sqrt{e^{-pt}|c_{00}|^2 + |c_{10}|^2}}. \tag{8.14}$$

Thus the pure no-mutation state $|0\rangle$ is stationary with respect to the continuous evolutionary drift. This state can always jump to the mutation state $|1\rangle$ and the probability rate of evolutionary jumps is constant, p.

If the input (pure) state differs from the pure no-mutation state, then it drifts towards the pure mutation state as given by (8.14) and the probability rate of sudden jumps to the pure mutation state $|1\rangle$ decreases as

$$P_{\text{jump}} = \frac{e^{-pt}p|c_{00}|^2\delta t}{e^{-pt}|c_{00}|^2 + |c_{10}|^2}.$$

Thus the evolution corresponding to the very simple environment operator given by (8.12) has the complex branching structure combining evolutionary jumps with continuous drifts.

The evolutions similar to one driven by the operator (8.12) are widely used in quantum physics. For example, take

$$L = \sqrt{p}|0\rangle\langle 1|. \tag{8.15}$$

It describes the *spontaneous emission* of a photon by an electron in the excited state $|1\rangle$, as a result, this electron jumps to the ground state $|0\rangle$. We remark that by the conventional interpretation of quantum mechanics, quantum randomness is *irreducible,* i.e., it is in principle impossible to find causal sources leading to a quantum jump. Quantum jumps are considered as an intrinsic feature of nature; this feature is not refinable. One may speculate that this quantum ideology can be extended to evolutionary jumps in biological evolution.

It follows from the presented analysis that the Lindblad operator L, (8.12), can be easily reconstructed from the experimental data. It depends on a single real parameter p, which is nothing else than the rate of transition probability from the state $|0\rangle$, no mutation, to the state $|1\rangle$, mutation.

8.4 "Entanglement" of Epimutations in Genome

We start with the construction of QL-representation of the information state of epigenome expressing CEI of the chromatin-marking type. Consider a cell with genome consisting of m genes $g_1, \ldots, g_m$. Let us assign to each gene g all its possible epimutations (of the chromatin-marking type); we simply enumerate them by numbers[7]: $j_g = 1, \ldots, k_g$.

The state of all potential epimutations in the gene g is represented as the superposition

$$|\psi_g\rangle = \sum_j c_{g;j} |j_g\rangle, \tag{8.16}$$

where $\sum_j |c_{g;j}|^2 = 1$.

What is the meaning of this superposition from the biological viewpoint? Can a gene really be in superposition of a few different epimutations?

Although our model is operational and in principle we are not interested in such questions, we make a comment to clarify the coupling of operational and biological descriptions of this situation. A cell by itself "knows its epigenome" at each instant of time; so it is well aware which epimutations have taken place up to this instant of time. However, a biologist performing an experiment with cells does not know the situation inside an individual cell in such detail. And superposition is related to the uncertainty of the *observer's information.*

If epimutations in different genes are independent from each other, then the QL-state of cell's epigenome is represented as the tensor product of states $|\psi_g\rangle$:

$$|\psi_{\text{epi}}\rangle = |\psi_{g_1}\rangle \otimes \ldots \otimes |\psi_{g_m}\rangle. \tag{8.17}$$

However, in living cells, most of the genes/proteins are correlated somehow forming a big network system. Thus, usually one epimutation affects other genes. Hence the assumption of independent epimutations is nonbiological. Therefore we have to consider more general states describing the consistent epimutations of all genes in the genome of a cell. These are *entangled states* (Sect. 2.5.1), which are widely used in quantum information theory:

$$|\psi_{\text{epi}}\rangle = \sum_{j_1 \cdots j_m} c_{j_1 \ldots j_m} |j_{g_1} \cdots j_{g_m}\rangle, \tag{8.18}$$

where $|j_{g_1} \cdots j_{g_m}\rangle$ is a short notation for the tensor product of states of superpositions in various genes, $|j_{g_1} \cdots j_{g_m}\rangle \equiv |j_{g_1}\rangle \otimes \ldots \otimes |j_{g_m}\rangle$ and the sum of all squared coefficients is equal to 1.

---

[7] Depending on the biological context, it is always possible to select a few epimutations of the main importance. Hence, the number k_g need not be very large. We state again that our model is operational. It need not be very detailed.

8.4.1 Interpretation of Entanglement in Genome

We remark that the notion of entanglement is in the very heart of quantum mechanics. However, although it is widely used in quantum information, the understanding of the physical essence of entanglement is far from to be complete, see Sect. 9.2. Nevertheless, there is complete consensus that entanglement implies *correlations*- in our epigenetic modeling these are correlations between epimutations in different genes. (Whether entanglement is simply the tensor product encoding correlations is an open question, see Sect. 9.2.)

Typically, in quantum foundations experts emphasize that correlations corresponding to an entangled state are "superstrong", i.e., they have amplitudes exceeding the amplitudes, which are possible for classical correlations. Experimentally, nonclassicality of correlations is demonstrated in the form of violation of Bell's inequality (Sects. 9.1 and 9.2). However for our modeling of epigenetic mutations, the debate on nonclassicality and superstrength of entanglement-correlations is not important.

The key point of the application of entanglement in modeling of epimutations is its *"nonlocal feature"*. We remark that the notion of quantum nonlocality is often used vaguely. Typically, especially in the relation to Bell's tests, nonlocality is understood as nonlocality of hidden variables (Sect. 9.2). In our cell biological studies we cannot assume such "real physical nonlocality" of variables describing gene expressions. We do not appeal to physical quantum effects in a cell. We state again that the origin of QL-representation is the uncertainty in information about cell behavior in experiment.

We shall appeal to another sort of nonlocality, which may be called *operational nonlocality*. The form of the tensor space representation (8.18) of *potential epimutations* in cell genome implies that epimutation in one gene implies consistent epimutations in other genes. If the state (8.18) is not factorized, then by changing of the environment of one gene, say g_1, and inducing, see Sect. 8.3, some epimutation in it, we can induce consistent epimutations in other genes. In the operational approach (endowed with the information interpretation) this global change of the genome state happens in *information space* and not in physical space, cf. [20, 21]. This is simply the update of information, which experimenter has to make by taking into account changing of the state of the gene g_1. Since information is encoded in probabilities, this global state updating is nothing else than an updating of probabilities. Such an updating is in general non-Bayesian, see also Chap. 4.[8]

---

[8] What is beyond such a global ("nonlocal") structure of state update? This is the separate question that can be ignored in the operational approach. However, biologists may want to have some picture of what happens beyond the operational description. Our picture is that gene expressions in genome are strongly correlated; these correlations can be nonlocal in the purely classical field manner: strongly correlated waves, including electromagnetic and chemical waves, in a cell. However, we state again that we shall proceed in the purely operational framework. Hence this classical wave picture for QL entanglement, see [22–27] need not be taken into account (and, moreover, it may be wrong). We state again that a model of brain functioning based on the representation of information by purely classical electromagnetic waves was elaborated in [28–30].

8.4.2 Environment Driven Quantum(-Like) "Computations" in Genome

Quantum nonlocality related to the tensor product structure of the representation of states of compound systems[9] is the main source of the speedup of quantum computers. However we do not advertise a rather common viewpoint that biological quantum computing plays some role in genetics and brain functioning. Quantum algorithms are based on *unitary dynamics* described by Schrödinger's equation. In our opinion such dynamics cannot survive on the biological scales of space, time, and temperature. In our QL-model a cell is an open QL-system; its dynamics is described by the quantum master equation; it is nonunitary. In particular, quantum entropy is not preserved (Sect. 8.3.1).

Nevertheless, in our QL-model we also explore quantum nonlocality (in the aforementioned operational interpretation) to speed up the epigenetic evolution in a living cell. Otherwise, i.e., by using the purely neo-Darwinian approach (first random (epi-)mutations and then selection), we would not be able to explain the high speed of the epigenetic evolution. If epimutations inducing new levels of gene expressions were randomly and independently generated and then selected, the evolution (in the case of epimutations in a large number genes as the reaction to the environment) would be too slow. Thus a QL-model without exploring entanglement, i.e., based only on superposition, would not match with experimental data.

Let an environment act on genes $g_1, \ldots, g_m$. For example, suppose that for g_1, as an individual gene, some epimutation, say M_{g_1}, can be useful in this environment. However, this epimutation may disturb functioning of other genes in a negative way. Hence, epimutations $M_{g_1}, \ldots, M_{g_n}$ induced by the environment have to be consistent. How can they become consistent? Either via iterations, first the state of epimutations $(M_{g_1}, \ldots, M_{g_n})$ is created, but the cell "feels" disagreement between levels of genes expressions corresponding to these epimutations. New epimutations are induced by this inconsistency and so on. This process is similar to Darwinian natural selection and approaching to consistency in genes expressions would take too long a period (for the time scale of one living cell).

Our proposal is that dynamics are entangled, and at one step all genes epimutate consistently. We state again that the main difference from quantum computing is using nonunitary evolution described by the quantum master equation instead of the unitary (Schrödinger) evolution. Hence, we use entanglement, but without unitary evolution. We can call such an approach *open quantum system computing*.

---

[9] We state once again that it has nothing to do with nonlocality of hidden variables.

8.4.3 Comparison with Waddington's Canalization Model

Our model based on QL-control by the environment of epigenetic evolution in combination with entanglement between epimutations in different genes matches very well with the epigenetic canalization model discussed by Sollars et al. [31], p. 73:

> In light of our data, we propose a refinement of the 1942 evolutionary E-canalization model of Waddington to an E-epigenetic canalizationE- model. In the canalization model, Waddington [32], environmental stress induces a novel phenotype, and selection of existing genetic variation in subsequent generations allows fixation of the novel phenotype. According to Waddington, "by such a series of steps, then, it is possible that an adaptive response can be fixed without waiting for the occurrence of a mutation which, in the original genetic background, mimics the response well enough to enjoy a selective advantage" [32]. In our epigenetic canalization model, we propose that an environmental stress causes a reduction in Hsp90 levels and, through some unknown interaction with TrxG proteins, induces an immediate E-chromatin effectE-. Our model allows an adaptive response to be E-fixedE- epigenetically, and therefore obviates the need to wait for the selection of existing genetic variation. In other words, it predicts a more rapid evolutionary process than is required for selection of existing genetic variation.

In Sect. 8.2.1 we pointed that, although under the environment driven QL-dynamics the superposition will be finally resolved and a steady state solution will be approached, the complete stabilization is possible only in the limit $t \to \infty$. Hence, for any finite interval of time, the total stabilization is impossible. This feature of our QL-model also matches well with observation of Sollars et al. [31], p. 73:

> Because of the inherent instability of epigenetic inheritance, fixation of an epigenetically-determined phenotype is probably less stable than fixation through a genetic selection mechanism. Waddington, for example, was unable to reduce the frequency of the crossveinless phenotype in negative selection experiments once the phenotype was fixed [33]. In contrast, after only two or three generations of negative selection, we observed a complete reversion to wild-type frequency of ectopic outgrowth in our sensitized iso-KrIf-1 strain in the geldanamycin selection experiment (data not shown). Similarly, epigenetic traits such as color variegation or cold adaptation in plants are unstably inherited, Bender [34] Kohler and Grossniklaus [36]. Therefore, a combination of both epigenetic and genetic mechanisms is probably required to explain the rapid changes in body plans that are observed in the fossil record, Gould and Eldredge [37].

8.5 Adaptive Dynamics in Space of Epigenetic Markers

We now proceed by operating with epigenetic markers as information quantities, i.e., without coupling each of them with a special form of cellular material. We enumerate all possible epigenetic markers involved in the process of evolution under the pressure of some fixed environment, $j = 1, \ldots, n$. Each marker can be quantified by the classical random variable $\xi_j = 1$, if this marker is created and then inherited, otherwise $\xi_j = 0$. These are observables that can be measured in experiments. The space of all classical states of the epigenome consists of vectors corresponding to the fixation of the values of all epigenetic markers: $\alpha = (\alpha_1, \ldots, \alpha_n)$, where $\alpha_j = 0, 1$.

This classical state space consists of 2^n points. This space is the basis of the classical information description of the process of epigenetic evolution. However, we move to the quantum information description by assuming that classical states can form superpositions. To match with the Dirac ket-vector notation, which is used in quantum physics, we denote the classical state α as $|\alpha\rangle \equiv |\alpha_1\rangle \ldots \alpha_n\rangle$. Then QL state space of (possible) epigenetic mutations, $\mathcal{H}_{\text{epi}}$, consists of superpositions of the form

$$|\psi\rangle = \sum_\alpha c_\alpha |\alpha\rangle,$$

where $\sum_\alpha |c_\alpha|^2 = 1$. This is a complex Hilbert space of dimension 2^n. Now we repeat our previous considerations, see Sects. 8.3 and 8.4, for epimutations of the chromatin-marking type. The QL adaptive dynamics described by the quantum master equation can be considered as a mixture of neo-Darwinian, neo-Lamarckian, and Wrightean evolutions. This cocktail of stochasticity and determinism is consistently represented in the QL operational framework. The final steady state gives to experimenters the classical probability distribution of the inherited epigenetic markers.

As we have seen in Sect. 8.4, entanglement plays a crucial role in the speedup of the epigenetic evolution. Since epimutations of the chromatin-marking type can be coupled to physical carriers, it was easy to use the standard notion of entanglement (as entanglement of systems) in the epigenetic framework. In general, epigenetic markers are merely information structures in a cell such as, e.g., self-sustaining regulatory loops. However, we are lucky, since recently a new general viewpoint on entanglement has been elaborated in quantum information community (Sect. 2.5.2). Entanglement can be considered not from the system viewpoint, but from the observer viewpoint as well. One considers a family of algebras of observables, say $\{\mathcal{A}_i\}$, on the total state space, in our case on $\mathcal{H}_{\text{epi}}$. Under some restrictions on these algebras the state space can be represented as the tensor product of subspaces corresponding to these algebras. (This is an algebraic representation of contextuality of some observables for a single cell, Sect. 2.5.2.) In our case we consider algebras of observables corresponding to different epigenetic markers, the corresponding subspaces are two dimensional qubit spaces, $\mathcal{H}_{\text{epi}} = \otimes_{j=1}^n \mathcal{H}_{j;\text{qubit}}$. Now we can use the notion of entanglement corresponding to this tensor product decomposition of the state space and repeat the speedup argument which was discussed in detail in Sect. 8.4.

References

1. Jablonka, E., Raz, G.: Transgenerational epigenetic inheritance: prevalence, mechanisms, and implications for the study of heredity and evolution. Q. Rev. Biol. **84**, 131–176 (2009)
2. Asano, M., Ohya, M., Khrennikov, A.: Quantum-like model for decision making process in two players game. Found. Phys. **41**(3), 538–548 (2010)
3. Asano, M., Ohya, M., Tanaka, Y., Khrennikov, A., Basieva, I.: On application of Gorini-Kossakowski-Sudarshan-Lindblad equation in cognitive psychology. Open. Syst. Inf. Dyn. **17**, 1–15 (2010)

4. Asano, M., Ohya, M., Tanaka, Y., Basieva, I., Khrennikov, A.: Dynamics of entropy in quantum-like model of decision making. J. Theor. Biol. **281**, 56–64 (2011)
5. Ogryzko, V.V.: A quantum-theoretical approach to the phenomenon of directed mutations in bacteria (hypothesis). Biosystems **43**, 83–95 (1997)
6. Ogryzko, V.V.: On two quantum approaches to adaptive mutations in bacteria. http://arxiv.org/ftp/arxiv/papers/0805/0805.4316.pdf
7. McFadden, J.J., Al-Khalili, J.V.: A quantum mechanical model of adaptive mutation. Biosystems **50**, 203–211 (1999)
8. McFadden, J.: Quantum Evolution. Harper (2000)
9. Donald, M.J.: A Review of Quantum Evolution. (2001) http://arxiv.org/abs/quant-ph/0101019
10. Fuchs, C. A.: Quantum mechanics as quantum information (and only a little more). In: Khrennikov, A. (ed.), Quantum Theory: Reconsideration of Foundations, Ser. Math. Modeling 2, Växjö University Press, Växjö, pp. 463–543 (2002)
11. Caves, C.M., Fuchs, C.A., Schack, R.: Quantum probabilities as Bayesian probabilities. Phys. Rev. A. **65**, 022–305 (2002)
12. Fuchs, ChA, Schack, R.: A quantum-Bayesian route to quantum-state space. Found. Phys. **41**, 345–356 (2011)
13. Zeilinger, A.: Dance of the Photons: From Einstein to Quantum Teleportation. Farrar, Straus and Giroux, New York (2010)
14. Balazs, A.: Some introductory formalizations on the affine Hilbert spaces model of the origin of life. I. On quantum mechanical measurement and the origin of the genetic code: a general physical framework theory. Biosystems **85**, 114–125 (2006)
15. Karafyllidis, I.: Quantum mechanical model for information transfer from DNA to protein. Biosystems **93**, 191–198 (2008)
16. Ingarden, R.S., Kossakowski, A., Ohya, M.: Information Dynamics and Open Systems: Classical and Quantum Approach. Kluwer, Dordrecht (1997)
17. Ohya. M., Volovich I.V.: Mathematical Foundations of Quantum Information and Computation and Its Applications to Nano- and Bio-systems, Springer (2011)
18. Rooney, P., Bloch, A., Rangan, C.H.: Decoherence control and purification of two-dimensional quantum density matrices under Lindblad dissipation. (2012) http://arxiv.org/abs/1201.0399
19. Fisher, R.A.: The Genetical Theory of Natural Selection. Clarendon Press, Oxford (1930)
20. Khrennikov, A.: On quantum-like probabilistic structure of mental information. Open. Syst. Inf. Dyn. **11**, 267–275 (2004)
21. Khrennikov, A.: Quantum-like brain: Interference of minds. Biosystems **84**, 225–241 (2006)
22. Acacio de Barros, J., Suppes, P.: Quantum mechanics, interference, and the brain. Math. Psychology 53, 306–313 (2009)
23. Khrennikov, A.: Detection model based on representation of quantum particles by classical random fields: Born's rule and beyond. Found. Phys. **39**, 997–1022 (2009)
24. Khrennikov, A., Ohya, M., Watanabe, N.: Classical signal model for quantum channels. J. Russ. Laser Res. **31**, 462–468 (2010)
25. Khrennikov, A.: Prequantum classical statistical field theory: Schrödinger dynamics of entangled systems as a classical stochastic process. Found. Phys. **41**, 317–329 (2011)
26. Khrennikov, A.: Description of composite quantum systems by means of classical random fields. Found. Phys. **40**, 1051–1064 (2010)
27. Khrennikov, A., Ohya, M., Watanabe, N.: Quantum probability from classical signal theory. Int. J. Quantum Inf. **9**, 281–292 (2011)
28. Khrennikov, A.: On the physical basis of theory of "mental waves". Neuroquantology **8**, S71–S80 (2010)
29. Khrennikov, A.: Quantum-like model of processing of information in the brain based on classical electromagnetic field. Biosystems **105**(3), 250–262 (2011)
30. Haven, E., Khrennikov, A.: Quantum Soc. Sci. Cambridge Press, Cambridge (2012)
31. Sollars, V., Lu, X., Xiao, L., Wang, X., Garfinkel, M.D., Ruden, D.M.: Evidence for an epigenetic mechanism by which HSP90 acts as a capacitor for morphological evolution. Nat. Genet. **33**, 70–74 (2003)

32. Waddington, C.H.: Canalization of development and the inheritance of acquired characters. Nature **150**, 563–565 (1942)
33. Waddington, C.H.: Genetic assimilation of an acquired character. Evolution **7**, 118–126 (1953)
34. Bender, J.: Plant epigenetics. Curr. Biol. **12**, R412–R414 (2002)
35. Koonin, E. V., Wolf, Yu. I.: Evolution of microbes and viruses: A paradigm shift in evolutionary biology? Frontiers in Cellular and Infection Microbiology. **119**2 (2012)
36. Kohler, C., Grossniklaus, U.: Epigenetics: the flowers that come in from the cold. Curr. Biol. **12**(4), R129–R131 (2002)
37. Gould, S.J., Eldredge, N.: Punctuated equilibrium comes of age. Nature **366**, 223–227 (1993)

Chapter 9
Foundational Problems of Quantum Mechanics

Abstract In this chapter, we briefly discuss the main foundational problems of quantum mechanics, especially the problem of hidden variables—variables which might provide finer description of quantum phenomena than given by quantum states.

Keywords Interpretations of quantum state · Entanglement · Bell inequality · Realism · Nonlocality · Hidden variables in biology and cognition · Biological potentiality

The present situation in quantum foundations is characterized by huge diversity of interpretations of the mathematical formalism of quantum mechanics and heavy debates on the interrelation of classical and quantum randomness, e.g., [1–8]. These debates have various philosophical dimensions and we shall discuss them briefly in this chapter. We tried to postpone such a discussion as long as possible, but the last chapter on evolution really has to be completed by deeper consideration of the philosophical aspects of usage of the quantum formalism in biology.

9.1 Bell's Inequality

Bell's inequality is an inequality [2] for a special linear combination of correlations for a few pairs of classical random variables. For special quantum states (entangled states), this inequality is violated. And this is predicted by the theoretical formalism of quantum mechanics. According to Bell [2], to explain its violation, one has to reject either realism (objectivity), i.e., a possibility to assign the values of observables to a system before measurement, or locality, i.e., instantaneous action at a distance. Personally, Bell (following Bohm) supported using of nonlocal classical (realistic) models beyond quantum mechanics. As was first found by I. Pitowsky, "Bell's inequality" has been known in mathematics since 19th century. It was derived by J. Boole (the inventor of Boolean logics), see [4] for details and the original source [9]. However, Boole used this equality to check whether a few random variables ("observables") can be described by single probability space. (He defined probability on a Boolean algebra. So the question was about a possibility to find a single Boolean

M. Asano et al., *Quantum Adaptivity in Biology: From Genetics to Cognition*,
DOI 10.1007/978-94-017-9819-8_9

algebra and probability on it.) If such probability space cannot be constructed, then the "Boole-Bell inequality" can be in principle violated. Later a Soviet probabilist Vorobj'ev found the most general conditions for the existence of a single probability space for a family of random variables, see again [4, 10] for details. Thus mathematicians considered Bell's type inequalities without any relation to nonlocality.

This viewpoint matches well with our discussion on non-Kolmogorovness of quantum probability, see Sect. 4.1.1. In Sect. 9.2 we shall show that a few sources of non-Kolmogorovness of collected statistical data can be associated with the real experimental situation, once again [4] for details. Hence, addiction to nonlocality, which is rather common in the quantum community, is not justified by the real experimental situation. At the same time "death of realism", nonobjectivity of quantum observables (see Sect. 9.3), mathematically can be coupled to non-Kolmogorovness.

It is clear from the above presentation that Bell's inequality can be used as a statistical test of non-Kolmogorovness (and in this sense quantum-likeliness) of data collected for pairs of observables of any origin, e.g., for biological observables. The essence of this test is that observables can be measured pairwise, but it is impossible (due to the specialty of experimental contexts) to measure these observables jointly. Thus the experimenter gets a collection of pairwise probability distributions, but not the total joint probability distribution. If the latter does not exist, one can expect the violation of Bell's inequality. Outside of physics the first Bell's test was done for recognition of ambiguous figures, Conte et al. [11]; for Bell's test for semantic models see Bruza et al. [12], see also the monograph of Bussemeyer and Bruza [13]. General analysis of usage of Bell's tests in cognitive psychology was performed by Dzhafarov et al. [14]. In cellular biology no Bell's type tests have yet been performed. And to find the corresponding experimental context and perform such an experiment is an exciting problem.

The LG-inequality, see Sect. 6.4.2, is also an inequality of the Bell-type. Thus our experimental test for it can be considered as one of the first Bell's type tests performed outside of physics.

9.2 Debate on Hidden Variables and Its Biological Dimension

This section is written for those who have already heard about the problem of hidden variables, violation of Bell's inequality, quantum nonlocality, (in)completeness of quantum mechanics. We discuss interpretation problems related to these topics. Thus, this section is written for biologists who may be afraid to use the quantum formalism in biology as a consequence of the aforementioned foundational problems and who are aware of these problems.

The debate on a possibility to go beyond the operational quantum formalism and to introduce hidden (i.e., yet unknown) variables, which determine the results of measurements, via the functional relation (1.1) was started about 90 years ago, by

Einstein and Bohr. Einstein was sure that sooner or latter a prequantum model will be created and the quantum formalism would be considered as simply an operational formalism appeared as a result of ignorance of the intrinsic degrees of freedom of quantum systems. Bohr (as well as Heisenberg, Fock, Landau, von Neumann,...) were sure that it is meaningless to look for such variables, since these are metaphysical. Bohr presented motivations (based on the existence of the indivisible Planck quant of action) that one cannot make better measurements than those described by the quantum formalism. Hence, intrinsic variables, even if one were able to find them, would be unmeasurable (metaphysical).

Later this position of fathers of quantum mechanics was strengthened, there were "proven" various no-go theorems. The most known are "theorems" of von Neumann [15], Bell [2], Kochen-Specker [16].[1] By these theorems any attempt to introduce hidden variables and to go beyond the operational quantum formalism would imply either contradiction with the predictions of quantum mechanics or such a pathological feature of hidden variables as nonlocality (Bell's theorem [2]). We remark that for Bohr and Heisenberg, such theorems had no real physical value, since they were sure that Heisenberg's uncertainty principle is incompatible with the existence of a subquantum classical structure. The first no-go theorem was proved by von Neumann in 1933 [15]. It was totally ignored by Bohr and Heisenberg. They just did not need it.

In fact, the no-go statements are dangerous pitfalls on the way towards applications of the quantum operational formalism in biology. (This problem is simply ignored by the majority of researchers, who formally apply the mathematical formalism of quantum mechanics, e.g., in psychology or cognitive science.) In biology intrinsic variables definitely exist; e.g., the levels of genes expressions in systems biological models of cell functioning or the neuronal variables in the brain science. It has been done a lot with the aid of such variables, e.g., in systems biology. And we cannot buy the idea of *biological nonlocality*, e.g., we cannot believe that human beings can be coupled on the mental level in the absence of physical interactions. It seems that we have to give up using the quantum operational formalism in biology. It is a powerful operational formalism describing measurements. However, its usage contradicts the possibility to introduce intrinsic variables in biological systems.

First of all, we note to the readers who are aware of the debate on hidden variables in quantum theory (in particular, Bell's inequality, quantum nonlocality, and so on) that the standard presentation of the contemporary state of art is not fair at all. The readers who are not experts in quantum foundations may get the impression that the problem was completely clarified and any model with local hidden variables would contradict the predictions of quantum mechanics. This is definitely not the case. The experimental setup of tests for violation of Bell's inequality involves a number of "loopholes" which provide possibilities to violate Bell's inequality and, moreover, to obtain quantum correlations for purely classical models. In fact, these loopholes are

[1] These statements are not really theorems in the rigorous mathematical sense. They are statements of the physical nature. Apart from mathematical assumptions (which are typically vaguely formulated), they contain a list of physical assumptions.

not simply experimental "technicalities", as it is typically presented in modern texts on the problem of hidden variables. As was emphasized in [4, 17], these are fundamental issues coupling the Bell's no-go argument with the real physical situation.

The most known loophole is the *detectors inefficiency loophole*. All experiments in which measurement stations are located on distances sufficient to check the assumption of locality are done in quantum optics. Detectors inefficiency is a severe problem of these experiments. It can imply *unfair sampling* [18]: Selection of different settings of measurement stations can make unfair cutoffs of the sample of photons emitted by a source. Statistical properties of sub-samples may depend on these settings. By using the probabilistic language we can say that different settings can induce different Kolmogorov probability spaces; so the total experimental framework involving a number of different experimental settings becomes (in general) non-Kolmogorovian. In such a framework, Bell's inequality can be easily violated and quantum correlations can be reproduced by simple classical models operating with a collection of incompatible Kolmogorov spaces [18].

Typically, this problem was treated as an experimental technicality, which will soon be clarified by the creation of photo-detectors with 100 % efficiency. A few years ago the leading groups in quantum optics got such detectors, *Superconducting Transition-Edge Sensors* (W-TESs)—the ultra-sensitive microcalorimeters. However, up to now all attempts to close the detection inefficiency loophole by using W-TESs have not been successful. Resolution of one "technicality", the inefficiency of detectors, induced another technicality: impossibility to couple, without essential losses, beams of entangled photons with microcalorimeters cooled to extremely low temperature. Two world's leading groups, one in Austria and another in USA, still work hard on this project. However, it seems that the chance of the success is very low.

Another loophole, which has attracted a lot of attention in the last years, is the *time window loophole*. Two entangled photons have to be detected at two stations. Under the condition of matching of distances from the source to measurement stations, in principle they have to be detected simultaneously. However in reality it is impossible. Therefore the experimenter selects a time window and identifies clicks of two detectors belonging to the same window. Of course, in this setup some pairs of entangled photons are discarded and, moreover, clicks belonging to different pairs can be matched, as well as clicks for photons from the source can be matched with noisy photons. This is also unfair sampling and probabilistically non-Kolmogorovness. In this framework, Bell's inequality can be easily violated.

Instead of photons, experimenters can use massive particles, e.g., ions or electrons. However, in this case it is very difficult to close "locality loophole", i.e., to separate measurement stations to a distance that would allow one to exclude the usual relativistic (with the velocity of light) communication. The authors of [17] even formulated a kind of complementarity principle for the detectors inefficiency and locality loopholes. By approaching to closing of one of them the experimenter would make the experimental situation essentially worse for closing another.

We summarize this discussion: *The real experimental situation for Bell's tests differs essentially from its presentation in standard textbooks on quantum mechanics. In the presence of various loopholes (which are not just* technicalities*) very simple* *classical probabilistic models can violate Bell's inequality and reproduce quantum correlations, which are typically considered as irreducible to classical correlations. Of course, it is a separate question whether such models with hidden variables describe the real physical situation.*

However, such models with hidden variables may match well with biological phenomena. Hence, there is a wide field for studies in foundations of quantum bio-informatics.

Another "do not be afraid of no-go theorems" argument was presented by Acacio de Barros [19]. He rightly pointed that the size of a biological system (too small) and the speed of information processing in it (too slow) are such that the discussion on the problem of a superluminal action on a distance, quantum nonlocality, is simply meaningless; see also his work [20] on QL-modeling of cognition.

Finally, we point to a recently developed prequantum model based on representation of quantum systems by random waves, *prequantum classical statistical field theory* [21–25], cf. [26]. This model in combination with the measurement theory based on interaction of random waves with detectors of the *threshold type* [27] produces correlations coinciding with quantum correlations. Here detectors can have 100 % efficiency; cf. Suppes and Acacio de Barros [28]. Applications of this approach to the creation of QL-representation of brain functioning based on classical electromagnetic random waves in the brain were presented in works [29–31]. However, we point out that in this book we do not advertise any particular model of physical realization of "brain's hardware". We proceed in the operational QL approach and we are interested, so to say, in "brain's software."

9.3 Nonobjectivity Versus Potentiality

Mathematical representation of uncertainty in bio-systems is the main topic of this book. We have emphasized repeatedly that quantum(-like) representation of uncertainty is not reduced to the classical probabilistic representation based on a single Kolmogorov probability space. QL-representation is more general. From the probabilistic viewpoint it is multi-(Kolmogorov)space representation.

In quantum foundations this discussion has the following philosophical dimension. In classical statistical mechanics of particles it is assumed that it is possible to assign to a system the values of observables, e.g., position, momentum, energy, before measurement. It is said that measured properties are *objective*, i.e., they are properties of an object. The common interpretation of violation of Bell's inequality is that in quantum physics this assumption is questionable (as well as the assumption of locality), Sect. 9.2.

Diversity definitely corresponds to the situation handled in classical statistical mechanics of particles; diversity can be considered as variability of objective, i.e., firmly established, properties of systems in a large ensemble (in theory infinitely large). Potentiality can be treated as a sort of nonobjectivity. It is, of course, very special kind of nonobjectivity, since, although the values of quantum observables are not predetermined, there is a kind of probabilistic predetermination on the level of probability distributions. Hence, nonobjectivity on the individual system level generates objectivity of statistical properties.[2] In this sense potentiality-nonobjectivity can be considered as an extension of the classical diversity-objectivity. Moreover, for the same system (depending on context) some properties can be predetermined before experiment while others cannot.

In quantum foundations the aforementioned picture is shadowed by the debate on locality/nonlocality, see Sect. 9.2. We definitely want to abandon the latter and exclude bio-nonlocality from this book. We need not reject completely the idea of *bio-nonlocality*, i.e., a possibility that bio-systems can interact on huge distances on the purely informational level, i.e., without sending physical signals propagating with the speed of light in physical space-time. Bio-nonlocality is just unnecessary for the phenomena under consideration in this book. In paper [34], it was shown that the common interpretation of rejection of *local realism* as the result of violation of Bell's inequality can be essentially sharpened. Experiments on *quantum contextuality* (Sect. 2.5.2), i.e., violation of Bell's type inequality for incompatible (i.e., jointly nonmeasurable) degrees of freedom of a single system, e.g., photon or neutron, imply that objectivity (and not locality) has to be rejected.

Nonobjectivity of observables is a very natural feature of bio-science. Typically a bio-system does not prepare "answers to questions" imposed by an observable (e.g., an experimenter), by an environment or even by itself.[3] "Answers" are created as reactions to imposed questions. Therefore bio-observables are fundamentally nonobjective. This is the main source of uncertainty in bioscience. This sort of uncertainty is evidently different from mere diversity. Potentiality is a proper image of such uncertainty.

References

1. Feynman, R., Hibbs, A.: Quantum Mechanics and Path Integrals. McGraw-Hill, New York (1965)
2. Bell, J.: Speakable and Unspeakable in Quantum Mechanics. Cambridge University Press, Cambridge (1987)
3. Khrennikov, A.: Växjö interpretation-2003: realism of contexts. In: Khrennikov, A.Yu. (ed.) Quantum Theory: Reconsideration of Foundations-2. Series Mathematical Modeling, vol. 10, pp. 323–338. Växjö University Press, Växjö (2004)
4. Khrennikov, A.: Contextual Approach to Quantum Formalism. Springer, Berlin (2009)
5. Khrennikov, A.: Interpretations of Probability, 2nd edn. De Gruyter, Berlin (2009). (completed)
6. Fuchs, C.A.: Quantum mechanics as quantum information (and only a little more). In: Khrennikov, A. (ed.) Quantum Theory: Reconsideration of Foundations. Series Mathematical Modeling, vol. 2, pp. 463–543. Växjö University Press, Växjö (2002)

[2] However clear philosophical understanding of the process of transition from individual nonobjectivity to statistical objectivity is still absent, cf. [5, 32, 33].

[3] For the latter, we have already pointed out, Chap. 1, that, opposite to quantum *physics*, in quantum(-like) biology it is impossible to draw a sharp distinction between an observer and a system. Self-observation is an important component of state-monitoring performed by bio-systems.

7. Fuchs, C.: Delirium quantum (or, where I will take quantum mechanics if it will let me). In: Adenier, G., Fuchs, C., Khrennikov, A.Yu. (eds.) Foundations of Probability and Physics-3. Series Conference Proceedings, vol. 889, pp. 438–462. American Institute of Physics, Melville (2007)
8. Garola, C., Khrennikov, A.: Dialog on beyond quantum. In: Khrennikov, A. (ed.) Quantum Theory: Reconsideration of Foundations-5. Series Conference Proceedings, vol. 1232, pp. 355–359. American Institute of Physics, Melville (2010)
9. Boole, G.: An Investigation of the Laws of Thought. Dover Edition, New York (1958)
10. Vorob'ev, N.N.: Consistent Families of Measures and Their Extensions. Theory Probab. Appl. **7**, 147–162 (1962)
11. Conte, E., Khrennikov, A., Todarello, O., Federici, A., Zbilut, J.P.: A preliminary experimental verification on the possibility of Bell inequality violation in mental states. Neuroquantology **6**, 214–221 (2008)
12. Bruza, P., Kitto, K., Ramm, B., Sitbon, L., Blomberg, S., Song, D.: Quantum-like non-separability of concept combinations, emergent associates and abduction. Log. J. IGPL **20**(2), 455–457 (2010)
13. Busemeyer, J., Bruza, P.D.: Quantum Cognition and Decision. Cambridge University Press, Cambridge (2012)
14. Dzhafarov, E.N., Kujala, J.V.: Quantum entanglement and the issue of selective influences in psychology: an overview. Lect. Notes Comput. Sci. **7620**, 184–195 (2012)
15. Von Neuman, J.: Mathematical Foundations of Quantum Mechanics. Princeton University Press, Princeton (1955)
16. Kochen, S., Specker, E.: The problem of hidden variables in quantum mechanical systems. J. Math. Mech. **17**, 59–87 (1967)
17. Khrennikov, AYu., Volovich, I.V.: Local realism, contextualism and loopholes in Bell's experiments. In: Khrennikov, AYu. (ed.) Foundations of Probability and Physics-2. Series Mathematical Modeling, vol. 5, pp. 325–344. Växjö University Press, Växjö (2002)
18. Adenier, G., Khrennikov, A.Yu.: Is the fair sampling assumption supported by EPR experiments? Phys. B: At., Mol. Opt. Phys. **40**(1), 131–141 (2007)
19. Acacio de Barros, J.: Quantum-like model of behavioral response computation using neural oscillators. Biosystems **110**, 171–182 (2012)
20. Acacio de Barros, J.: Joint probabilities and quantum cognition. In: Proceedings of the International Conference on Quantum Theory: Reconsiderations of Foundations-6. Växjö, Sweden, 11–14 June (2012)
21. Khrennikov, A.: Detection model based on representation of quantum particles by classical random fields: Born's rule and beyond. Found. Phys. **39**, 997–1022 (2009)
22. Khrennikov, A., Ohya, M., Watanabe, N.: Classical signal model for quantum channels. J. Russ. Laser Res. **31**, 462–468 (2010)
23. Khrennikov, A.: Prequantum classical statistical field theory: Schrödinger dynamics of entangled systems as a classical stochastic process. Found. Phys. **41**, 317–329 (2011)
24. Khrennikov, A.: Description of composite quantum systems by means of classical random fields. Found. Phys. **40**, 1051–1064 (2010)
25. Khrennikov, A., Ohya, M., Watanabe, N.: Quantum probability from classical signal theory. Int. J. Quantum Inf. **9**, 281–292 (2011)
26. Ohya, M., Watanabe, N.: A new treatment of communication processes with Gaussian channels. Jpn. J. Ind. Appl. Math. **3**, 197–206 (1986)
27. Khrennikov, A.: Role of detectors and their proper calibration in inter-relation between classical and quantum optics. Opt. Eng. **51**, Article number 069001 (2012)
28. Acacio de Barros, J., Suppes, P.: Quantum mechanics, interference, and the brain. Math. Psychol. **53**, 306–313 (2009)
29. Khrennikov, A.: On the physical basis of theory of "mental waves". Neuroquantology **8**, S71–S80 (2010)
30. Khrennikov, A.: Quantum-like model of processing of information in the brain based on classical electromagnetic field. Biosystems **105**(3), 250–262 (2011)

31. Haven, E., Khrennikov, A.: Quantum Social Science. Cambridge Press, Cambridge (2012)
32. Plotnitsky, A.: Reading Bohr: Physics and Philosophy. Springer, Heidelberg (2006)
33. Plotnitsky, A.: On the reasonable and unreasonable effectiveness of mathematics in classical and quantum Physics. Found. Phys. **41**, 466–491 (2011)
34. Khrennikov, A.: Towards a resolution of dilemma: nonlocality or nonobjectivity? Int. J. Theor. Phys. **51**, 2488–2502 (2012)

Chapter 10
Decision and Intention Operators as Generalized Quantum Observables

Abstract Since the classical formula of total probability is violated for data from both QM and cognitive psychology and the former has developed the advanced mathematical apparatus for the study of the interference effect, it is natural to apply this apparatus to describe statistical data of the latter. However, there is one pitfall; one problem has to be taken into account and carefully analyzed. The experimentally obtained matrices of transition probabilities are not doubly stochastic as is not the case for the matrices corresponding to the transitions from one orthonormal basis to another (the bases of the eigenvectors of the Hermitian operators representing quantum observables). Therefore "cognitive observables" ("mental observables") cannot be represented by Hermitian operators (at least with nondegenerate spectra). A possibility to solve this problem, namely, to use lifting maps, was explored in the previous chapters. A lifting map describes in detail the interaction of a system with an environment; this map contains, as parameters, the state of environment, the operator of the interaction between the system and the environment, and the measurement device. However, sometimes such a detailed description of the environment can be reduced to mere representing observables of systems by positive operator valued measures. In this chapter we demonstrate the power of usage of positive operator valued measures (POVMs) in biological science on the example of quantum modelling of cognition.

Keywords Positive operator valued measure (POVM) · Quantum cognition · Mental observables · Degenerate and non-degenerate spectra · Lifting · Interference for observables represented by POVMs

As noted in Chap. 2, since the classical formula of total probability (FTP) is violated for data from both QM and cognitive psychology and the former has developed the advanced mathematical apparatus for the study of the interference effect, it is natural to apply this apparatus to describe statistical data of the latter. However, there is one pitfall; one problem has to be taken into account and carefully analyzed. The experimentally obtained matrices of transition probabilities are not *doubly stochastic* (Sect. 4.1.1) as is not the case for the matrices corresponding to the transitions from one orthonormal basis to another (the bases of the eigenvectors of the Hermitian operators representing quantum observables). Therefore "cognitive observables" ("mental observables") cannot be represented by Hermitian operators (at least with nondegenerate spectra). A possibility to solve this problem,

M. Asano et al., *Quantum Adaptivity in Biology: From Genetics to Cognition*,
DOI 10.1007/978-94-017-9819-8_10

namely, to use lifting maps, was explored in the previous chapters. A lifting map describes in detail the interaction of a system with an environment; this map contains, as parameters, the state of environment σ, the operator of the interaction between the system and the environment V, and the measurement device M. However, sometimes such a detailed description of the environment can be reduced to mere representing observables of systems by *positive operator valued measures* (POVM). For biological applications, it is important that matrices of transition probabilities of POVMs are in general *not doubly stochastic*. Therefore it is natural to apply POVMs to describe biological statistical data. In this chapter we present a QL-model of decision making based on the representation of the *decision and intention operators* by POVM.

Although the interrelation between violation of the classical FTP and quantum interference of probabilities is well studied, see Chap. 2, typically the situation is restricted to probabilities corresponding to conventional quantum observables and pure states (Hermitian operators and complex state vectors). The FTP with interference term for the POVM-observables and the mixed states (density operators) was derived in [1]. In this case the interference term cannot be parameterized just by the "interference angle" θ. It contains a few parameters admitting an interesting cognitive interpretation, see Sect. 10.3.

We model decision making in the two party games: one party, say Bob, has to make decision on some problem B. The payoff depends on the actions of another party, say Alice. Therefore Bob has to take into account Alice's intentions A, see Chap. 6. Bob's mental state is described by the density operator ρ, his decision operator is given by the POVM M^b; Bob also has to analyze possible Alice's actions, this "measurement" is represented by the POVM M^a. We stress that both the decision making operator M^b and the intention operator M^a represent *self-measurements,* see Chap. 1 for discussion.

10.1 Generalized Quantum Observables

We start with the presentation of the fundamentals of the theory of *positive operator valued measures* (POVM). (POVMs were already mentioned at the end of Sect. 1.3 and they were used in Sect. 4.4 to represent "event systems".)

10.1.1 POVMs

Let H be a finite dimensional complex Hilbert space endowed with the scalar product $\langle \cdot, \cdot \rangle$. The space of density operators is denoted by the symbol $\mathscr{D}(H)$.

Definition A positive operator valued measure (POVM) is a family of positive operators $\{M_j\}$ such that $\sum_{j=1}^{m} M_j = I$, where I is the unit operator.

(We considered the simplest case of a discrete operator valued measure on the set of indexes $J = \{1, 2, ..., m\}$.) It is convenient to use the following representation of POVMs:

$$M_j = V_j^\star V_j, \tag{10.1}$$

where $V_j : H \to H$ are linear operators.

A POVM can be considered as a random observable. Take any set of labels $\alpha_1, ..., \alpha_m$, e.g., for $m = 2, \alpha_1 =$ yes, $\alpha_2 =$ no. Then the corresponding observable takes these values (for systems in the state ρ) with the probabilities

$$p(\alpha_j) \equiv p_\rho(\alpha_j) = \mathrm{tr}\rho \mathrm{M_j} = \mathrm{tr} \mathrm{V_j} \rho \mathrm{V_j^\star}. \tag{10.2}$$

We are also interested in *post-measurement states*. Suppose that the state ρ is given, a generalized observable is measured, and the value α_j is obtained. Then the output state after this measurement has the form

$$\rho_j = \frac{V_j \rho V_j^\star}{\mathrm{tr} \mathrm{V_j} \rho \mathrm{V_j^\star}}. \tag{10.3}$$

10.1.2 Interference of Probabilities for Generalized Observables

Consider two generalized observables a and b corresponding to POVMs $M^a = \{V_j^\star V_j\}$ and $M^b = \{W_j^\star W_j\}$, where $V_j \equiv V(\alpha_j)$ and $W_j = W(\beta_j)$ correspond to the values α_j and β_j.

If there is given the state ρ, the probabilities of observations of values α_j and β_j have the form

$$p^a(\alpha) = \mathrm{tr}\rho \mathrm{M^a}(\alpha) = \mathrm{tr} \mathrm{V}(\alpha) \rho \mathrm{V}^\star(\alpha), \;\; \mathrm{p}(\beta) = \mathrm{tr}\rho \mathrm{M^b}(\beta) = \mathrm{tr} \mathrm{W}(\beta) \rho \mathrm{W}^\star(\beta). \tag{10.4}$$

Now we consider two consecutive measurements: first the a-measurement and then the b-measurement. If in the first measurement the value $a = \alpha$ was obtained, then the initial state ρ was transformed into the state

$$\rho_\alpha^a = \frac{V(\alpha) \rho V^\star(\alpha)}{\mathrm{tr} \mathrm{V}(\alpha) \rho \mathrm{V}^\star(\alpha)}. \tag{10.5}$$

For the consecutive b-measurement, the probability to obtain the value $b = \beta$ is given by

$$p(\beta|\alpha) = \mathrm{tr}\rho^{\mathrm{a}}(\alpha)\mathrm{M}^{\mathrm{b}}(\beta) = \frac{\mathrm{tr}\mathrm{W}(\beta)\mathrm{V}(\alpha)\rho\mathrm{V}^{\star}(\alpha)\mathrm{W}^{\star}(\beta)}{\mathrm{tr}\mathrm{V}(\alpha)\rho\mathrm{V}^{\star}(\alpha)}. \tag{10.6}$$

This is the conditional probability to obtain the result $b = \beta$ under the condition of the result $a = \alpha$.

We set

$$p(\alpha, \beta) = p^{a}(\alpha)p(\beta|\alpha). \tag{10.7}$$

This is the probability to obtain the result (α, β) in the consecutive measurement of a and then b. To find $p(\alpha, \beta)$ with the aid of the probability $p(\alpha)$ and the conditional probability $p(\beta|\alpha)$, we apply the classical Bayes formula. Thus "quantumness" is in the probabilities $p(\alpha)$ and $p(\beta|\alpha)$; the probability $p(\alpha, \beta)$ is the result of their classical combination.

We remark that the probability $p(\alpha, \beta)$ can be expressed directly in terms of POVMs:

$$p(\alpha, \beta) = \mathrm{tr}\rho\mathrm{V}^{\star}(\alpha)\mathrm{W}^{\star}(\beta)\mathrm{W}(\beta)\mathrm{V}(\alpha). \tag{10.8}$$

In general, the family of probabilities $p(\alpha, \beta)$ cannot be considered as the joint probability distribution of two classical random variables, the pair (a, b). In the same way we introduce the probability

$$p(\beta, \alpha) = p^{b}(\beta)p(\alpha|\beta). \tag{10.9}$$

This is the probability to obtain the result $b = \beta$ in the first measurement and then the result $a = \alpha$. One can easily find examples of POVMs (and even Hermitian operators) such that

$$p(\alpha, \beta) \neq p(\beta, \alpha). \tag{10.10}$$

We call $p(\alpha, \beta)$ and $p(\beta, \alpha)$ *ordered joint probabilities.*

10.2 Formula of Total Probability with the Interference Term for Generalized Quantum Observables and Mixed States

We recall (Chap. 2) that, for two classical random variables a and b, which can be represented in the Kolmogorov measure-theoretic approach, the formula of total probability (FTP) has the form

$$p^{b}(\beta) = \sum_{\alpha} p^{a}(\alpha)p(\beta|\alpha). \tag{10.11}$$

Further we restrict our consideration to the case of dichotomous variables, $\alpha = \alpha_1, \alpha_2$ and $\beta = \beta_1, \beta_2$.

We are now interested in the version of FTP with the interference term, see (4.10), for generally nonpure states given by density operators and generalized quantum observables given by two (dichotomous) PVOMs:

$$p^b(\beta) = p^a(\alpha_1)p(\beta|\alpha_1) + p^a(\alpha_2)p(\beta|\alpha_2) + 2E -_\beta \sqrt{p^a(\alpha_1)p(\beta|\alpha_1)p^a(\alpha_2)p(\beta|\alpha_2)}, \tag{10.12}$$

or by using the ordered joint probabilities

$$p^b(\beta) = p(\alpha_1, \beta) + p(\alpha_2, \beta) + 2\lambda_\beta\sqrt{p(\alpha_1, \beta)p(\alpha_2, \beta)}. \tag{10.13}$$

Here the coefficient of interference λ_β has the form:

$$\lambda_\beta = \frac{\sum_{i=1,2} \mathrm{tr}\rho\{W^\star(\beta)V^\star(\alpha_i)V(\alpha_i)W(\beta) - V^\star(\alpha_i)W^\star(\beta)W(\beta)V(\alpha_i)\}}{2\sqrt{p^a(\alpha_1)p(\beta|\alpha_1)p^a(\alpha_2)p(\beta|\alpha_2)}} \tag{10.14}$$

Introduce the parameter

$$\gamma_{\alpha\beta} = \frac{\mathrm{tr}\rho \mathrm{W}^\star(\beta)\mathrm{V}^\star(\alpha)\mathrm{V}(\alpha)\mathrm{W}(\beta)}{\mathrm{tr}\rho \mathrm{V}^\star(\alpha)\mathrm{W}^\star(\beta)\mathrm{W}(\beta)\mathrm{V}(\alpha)} = \frac{p(\beta, \alpha)}{p(\alpha, \beta)}. \tag{10.15}$$

This parameter is equal to the ratio of the ordered joint probabilities of the same outcome, but in the different order, namely, "*b* then *a*" or "*a* then *b*". Then [1]

$$\lambda_\beta = \frac{1}{2}\Big[\sqrt{\frac{p(\alpha_1, \beta)}{p(\alpha_2, \beta)}}(\gamma_{\alpha_1\beta} - 1) + \sqrt{\frac{p(\alpha_2, \beta)}{p(\alpha_1, \beta)}}(\gamma_{\alpha_2\beta} - 1)\Big]. \tag{10.16}$$

In principle, this coefficient can be larger than one [1]. Hence it cannot be represented as

$$\lambda_\beta = \cos\theta_\beta \tag{10.17}$$

for some angle ("phase") θ_β. However, if POVMs M^a and M^b are, in fact, spectral decompositions of Hermitian operators, then the coefficients of interference are always less than one, i.e., one can find phases θ_β.

The transition from the classical FTP to the FTP with the interference term (10.12) can be considered as an extension of the parameter space. Besides the probabilities for the results $a = \alpha$ and $b = \beta$ and the conditional probabilities $p(\beta|\alpha)$, new parameters λ_β, the coefficients of interference between observables a and b (for the state ρ), are taken in consideration.

We present an interpretation of the structure of the coefficients of interference. We start with terms $\gamma_{\alpha\beta}$. They express *noncommutativity of measurements* or nonclassicality of the joint probability distribution. For classical probability all coefficients $\gamma_{\alpha\beta} = 1$ (and hence all interference coefficients $\lambda_\beta = 0$).

We also introduce the coefficients

$$\mu_\beta = \frac{p(\alpha_1, \beta)}{p(\alpha_2, \beta)}. \tag{10.18}$$

They express the relative magnitude of probabilistic influences of the results $a = \alpha_1$ and $a = \alpha_2$, respectively, on the result $b = \beta$.

Thus the coefficients of interference are composed of the coefficients of noncommutativity, which are weighted with the relative magnitudes of influences:

$$\lambda_\beta = \frac{1}{2}[(\gamma_{\alpha_1\beta} - 1)\sqrt{\mu_\beta} + (\gamma_{\alpha_2\beta} - 1)/\sqrt{\mu_\beta}]. \tag{10.19}$$

10.3 Decision Making with Generalized Decision and Intention Operators

We now present again the schemes of decision making that are based on classical and QL FTP. In the latter case we discuss the role of "interference parameters" in QL FTP of Sect. 10.2.

10.3.1 Classical Scheme

There are two parties, Alice and Bob. They operate under some complex of conditions (physical, social, financial) C, context. Each can act only in two ways: $a = \alpha_1, \alpha_2$ and $b = \beta_1, \beta_2$. Bob's decisions depend crucially on Alice's actions. However, in general Bob does not know precisely which action $a = \alpha_1$ or $a = \alpha_2$ will be chosen by Alice. Therefore Bob can only guess. He estimates subjectively the probabilities $p^a(\alpha_1)$ and $p^a(\alpha_2)$ of her possible actions. He can also estimate probabilities of his own actions conditioned by Alice's actions, $p(\beta|\alpha)$. Finally, Bob estimates the probabilities $p^b(\beta_1)$ and $p^b(\beta_2)$ by using FTP (10.11). If, e.g., $p^b(\beta_1) > p^b(\beta_2)$, then Bob makes the decision $b = \beta_1$.

This scheme is realized in the brain on the unconscious level. Subjective estimates of probabilities for Alice's actions and conditional probabilities for his own actions conditioned by her actions typically are not present on the conscious level.

10.3.2 Quantum-Like Scheme

This is a generalization of classical scheme. Besides the aforementioned probabilities, Bob estimates the measure of incompatibility of his possible actions and possible Alice's actions, the coefficients of interference. The latter estimate is combined of the estimate of the effect of noncommutativity of Bob's and Alice's actions and relative magnitudes of probabilistic influences of Alice's actions on Bob's actions. To estimate $\gamma_{\alpha\beta}$, Bob has to estimate not only the ordered probability $p(\alpha, \beta)$ of his own action $b = \beta$ under the condition that Alice would act as $a = \alpha$, but also the ordered probability $p(\beta, \alpha)$ of possible Alice's action $a = \alpha$ under the condition that she assumes Bob's action $b = \beta$. However, since the absolute magnitudes of these probabilities are not important, Bob is interested only in their ratio.

Finally, Bob estimates probabilities of his actions by using the FTP with the interference term (10.12).

This scheme is also realized on the unconscious level. In particular, all the aforementioned measures of quantumness are estimated subjectively (on the basis of collected experience).

We can say that in the QL-scheme the counterfactual arguments play a crucial role. To estimate the coefficients of noncommutativity $\gamma_{\alpha\beta}$, Bob has to imagine how Alice would act if he acted as $b = \beta$. These probabilities are compared with the probabilities of Bob's own actions conditioned by possible Alice's actions. If Bob guesses that Alice makes the same estimation of the probabilities of his actions conditioned by her actions as for her own actions conditioned the corresponding Bob's actions, then $p(\alpha, \beta) = p(\beta, \alpha)$ and $\gamma_{\alpha\beta} = 1$. In this case the QL and classical FTP and, hence, the decision making schemes coincide. Thus *in this decision making scheme quantumness is related asymmetrically in the estimation of the action-reaction probabilities.* Bob can imagine that Alice is friendlier than him or that Alice is more defect-inclined (we consider the Prisoner's Dilemma game as the basic example of the situation under consideration).

In the operational QL-formalism all probabilities and components of the coefficients of interference are encoded in the Bob's mental state, the density operator ρ, his decision making observable, POVM M^b, and his observable for Alice's intentions, POVM M^a (We remark that these are self-observables).

Reference

1. Khrennikov, A., Loubenets, E.: On relation between probabilities in quantum and classical experiments. Found. Phys. **34**, 689–704 (2004)

Index

M. Asano et al., *Quantum Adaptivity in Biology: From Genetics to Cognition*,
DOI 10.1007/978-94-017-9819-8

MIX
Papier aus verantwortungsvollen Quellen
Paper from responsible sources
FSC® C105338

If you have any concerns about our products,
you can contact us on
ProductSafety@springernature.com

In case Publisher is established outside the EU,
the EU authorized representative is:
Springer Nature Customer Service Center GmbH
Europaplatz 3, 69115 Heidelberg, Germany

Printed by Libri Plureos GmbH
in Hamburg, Germany